Richard J. Dunglison

A new School Physiology and Hygiene

With Special Reference to the Action of Alcohol and Narcotics

Richard J. Dunglison

A new School Physiology and Hygiene
With Special Reference to the Action of Alcohol and Narcotics

ISBN/EAN: 9783744670760

Printed in Europe, USA, Canada, Australia, Japan

Cover: Foto ©berggeist007 / pixelio.de

More available books at **www.hansebooks.com**

A NEW

SCHOOL PHYSIOLOGY

AND

HYGIENE

WITH SPECIAL REFERENCE TO THE ACTION OF
ALCOHOL AND NARCOTICS

BY

RICHARD J. DUNGLISON, A.M., M.D.

AUTHOR OF "HANDBOOK OF DIAGNOSIS, THERAPEUTICS, ETC." EDITOR OF DUNGLISON'S
"MEDICAL DICTIONARY" AND "HISTORY OF MEDICINE." SECRETARY
OF THE AMERICAN ACADEMY OF MEDICINE, ETC.

CHICAGO   NEW YORK

THE WERNER COMPANY

SCHOOL PHYS.

PREFACE.

It has been the aim of the author in the following pages to impart such information on the interesting subject of Physiology as will make the reader familiar with the general structure of his own body. To young people this knowledge is especially important and desirable. They have the opportunity to be practically benefited by such instruction at a sufficiently early period in life to enable them to be watchful in regard to the laws of health, which they must necessarily study in connection with Physiology. Experience as a physician fully justifies the author in the statement that it is in ignorance of such laws, and in consequence of the lack of proper education in this useful branch, that so many adults are daily violating the simplest principles of health. No more important subject can be taught in the schools than that which instructs the student in the principles of his own formation. In the frequent allusion made throughout this work to Comparative Anatomy and Physiology—the anatomy and physiology of other animals compared with man—the reader will be able to appreciate the lofty position occupied by him in the scale of creation. The knowledge which will thus be acquired by him of the higher state of development characteristic of man, and of the possibilities of still greater perfection of his intellectual and physical powers under proper education and training, will be one of the most valuable and improving lessons which Physiology can teach.

RICHARD J. DUNGLISON, M. D.

CONTENTS.

1 *   5

6 CONTENTS.

ANIMAL HEAT.

SECRETION.

THE NERVOUS SYSTEM.

THE SENSES.

PHYSIOLOGY.

It was formerly the custom to divide all bodies into the three classes, Animals, Vegetables, and Minerals, and this mode of division is still adopted by some writers. Animals and vegetables resemble each other, however, in so many respects, especially in the possession of *organs*, that it has seemed desirable to make a different classification. The division now generally adopted is into *organized* bodies and *inorganic*, the organized being composed of organs for the performance of certain duties, while no such arrangement exists in the inorganic. Animals and vegetables, being made up of organs, belong to the class of organized bodies; minerals are inorganic.

Physiology proper.—Physiology is that branch of science which treats of the healthy exercise of the various organs of the body. The term is derived from two Greek words denoting *doctrine of nature*. *Vegetable Physiology* considers the properties of vegetables only, in all that concerns their life and nourishment. When the term Physiology is used alone, it is generally understood as restricted to man and animals. It is called *Human Physiology* when applied to man alone; it is known as *Comparative Physiology* when applied to animals generally as compared with man.

The chief subject of our inquiry will, of course, be the physiology of man, but allusion will frequently be made to comparative physiology, so that the place occupied by man

9

in the scale of animals may be thoroughly understood. In whatever department we study physiology, we cannot fail to be struck with the vastness of its scope, for it embraces a knowledge of the progress and decay of all forms of vegetable and animal life, from brute matter, which is devoid of vitality, to the most perfect animal—man himself—who possesses it in the highest vigor.

Anatomy.—Anatomy is a distinct branch of inquiry, the term being derived from a Greek word meaning *dissection*. As generally used it signifies that study by which, through the medium of dissection, we become acquainted with the structure of the various parts that enter into the composition of the bodies of animals. It is really necessary, as a general rule, to understand the structure of the different organs of the body—that is, their anatomy—before we can fully appreciate their physiology. We can sometimes infer from the anatomy of a part what are its uses, but some of the organs, such as the brain, give us but little information. We know that there can be no exercise of intellect without a brain, and yet when we come to examine that important organ we learn absolutely nothing from its structure. It is apparently only a mass of fat, phosphorus and other chemical materials arranged in a peculiar way and protected from injury in a bony covering.

This is all we learn from its anatomy or chemistry, although it is an organ capable of the highest duties and has many important organs under its control. Anatomically, the eye is a perfect organ and the structure of the tongue is very familiar to us, yet the nicest dissector has failed to show us how the eye possesses the power of giving us our ideas of external objects or the tongue can impart to us the hundreds of varied flavors.

Comparative Anatomy is the study of the structure of

other animals than man as compared with him. When we see differences in the arrangement of their organs we know that there must be some different duty performed, and thus we may acquire a knowledge of their physiology. For instance, we see a variation in the anatomical arrangement of the stomachs of animals, and this we find occurs in accordance with a variation in the kind of food they live upon and the amount of digestion required.

Aids to Physiology.—Even disease or malformation may aid us in learning physiology. Cases have occurred where life was carried on for a while without any brain, the child breathing and swallowing just as well without it, although it would not, of course, exhibit any mental power. This shows that the acts of respiration and of swallowing are not controlled by the brain, and so by the absence of an anatomical part we derive information negatively as to its duties.

The laws of *Physics* are also involved in the study of physiology. Sight or vision is an example of this, the eye being mainly a physical apparatus, a transparent body influenced by the laws of light. If the physical part be imperfect, vision is defective. So, too, the muscles are a physical apparatus, acting like levers of different kinds according to the angle or mode in which they are inserted into the bones. As we shall see hereafter, the arrangement of the ear for hearing is partly a physical apparatus.

Chemistry also aids us in the study of physiology. Digestion, for example, is mainly a chemical process, the juices in the stomach and small intestines which effect it being made up chiefly of acids and a ferment. The fluids in the mouth which act on the food in that cavity produce their effect by chemical action. Respiration or breathing is also a chemical and mechanical process, being an exchange of gases into and out of the lungs, oxygen being received into those organs, and carbonic acid being given off. These functions or processes are also, to some extent, affected or regulated by a nervous system. There are many functions that are performed wholly or in part in a mechanical or chemical manner, and their action is easily understood; yet the controlling power—the vital action—is wholly beyond our calculation. The healthy action of the heart, for instance, is dependent on the movement of its own muscles and of valves in its interior, which act in a purely mechanical manner; and yet we cannot explain minutely why such action is carried on through life and sometimes by artificial stimulation after death. We say it is a power existing within itself, under proper control, and there our explanation ends.

Differences between Organized and Inorganic Bodies.—Inorganic bodies, such as minerals, are not born, as animals are; they do not possess life, and have not a certain definite shape like animals and vegetables. In each of these last—the animal as well as the vegetable—there are a particular form and recognized size, which belong to that kind and to every part of it. Vegetables, for instance, have a certain relation of size and shape of their stems and roots and limbs; and animals have their bodies and all their parts framed on one general and uniform plan. The animal and vegetable are constantly undergoing change, while the mineral is at rest. As long as life lasts the organized body is never in perfect repose; it is continually receiving food in some shape or other with which to support life, or carrying on the circulation of blood or of some other fluid necessary to its existence, or performing some other duty assigned to its various organs. In the animal and vegetable the different parts have a certain dependence upon each other, and no portion can be detached from another without inflicting injury upon the whole body; but large pieces can be broken from a mineral, and its shape entirely changed by force applied to it, without any such sympathy on the part of the rest of the body. When we examine the structure of these different classes of bodies we find a great variation, for, as already stated, animals and vegetables have a regular uniform arrangement, and certain parts which serve as organs. These are known by definite names: in the vegetable, for instance, they are the roots, leaves, flowers, bark, etc.; in the animal, the nerves, muscles, blood-vessels, etc. In a mineral or inorganic substance the structure is alike throughout. In the animal or vegetable, as we shall hereafter see, the particles are arranged in the form of fibres which run in all directions through and across one another to form a tissue, through which the different organs are made up.

Organic bodies pass through a series of changes, which in the animal particularly are known as *ages*, during which the organs become developed, attain their full growth, and then decay. The condition of the inorganic body is very different; it only changes its color or its shape when affected by the influence of blows or external agents, as rain, etc. A stone may increase in size by the addition of earthy or other matter to its surface, but it never grows in the same sense that animals and vegetables are said to grow.

So, too, with regard to the *death* of either of these organized bodies; they generally cease to have life as the result of some cause operating from within, as when a vital organ—the heart or lungs, for example—becomes diseased and unable to carry on its duties. The inorganic body has no death; it may be broken to atoms by force,

2

or it may crumble into dust, but there is no cessation of life here, for actual life never existed. Some of the lower forms of animals and vegetables may exist but a short time, while man may sometimes live a century; but this short or long duration of life is controlled by the interior organization of the animal and by external causes operating upon it.

When the organs of the animal perform the work assigned to them properly and in full vigor, they are said to be *healthy*, and the general condition is said to be one of *health*. The reverse of this, when the action of the organs is interfered with from any cause, is called *disease*. The term *hygiene* has been applied to that branch of study which includes all the different methods, medical or otherwise, devised for the preservation of health; for Hygeia, in the ancient mythology, was the goddess of health.

Differences between Animals and Vegetables.—It is not so easy to draw the line here as it was between these two classes of organized bodies and the inorganic. Both animals and vegetables require nourishment, and hence we say that *nutrition* is a function or performance common to both. But animals have a sense of feeling—*sensation*—which vegetables do not have, and they also have the power of moving at their own pleasure, and this is called *voluntary motion*. Under this action of the will they are able to move any part of the body as they may wish, or to go from place to place in any direction they may desire. Some of the lowest forms of animal life seem little higher in the scale than the vegetables they resemble, but the physiologist has certain laws by which he is able to distinguish them. As a general rule, vegetables contain more solids than fluids; in the animal, as man, although externally, to the ordinary observer, largely solid, the fluids contained in the various tissues and organs of the body are greatly in excess. In the vegetable there is but one elementary tissue observable in its structure, a tissue made up of *areolæ*, or spaces, and hence called *areolar tissue.* The animal possesses this, and in addition other primary tissues, such as the *muscular* and *nervous tissues*, all of which will be described hereafter. The vegetable, having no powers of sensation or feeling, or voluntary motion, does not require a brain or nervous system, or even muscles which either a brain or nervous system can control.

When we say that animals and vegetables present these general differences in their structure, we must not forget that in the lowest forms of animal and vegetable life such distinctions are not perceptible. Some vegetables of this kind contain more fluids than solids, and in some of the minute animals there is a total absence of muscles and nerves, and even of organs, as heart, lungs, or stomach, and of vessels to convey the fluids from one part of the body to another. When flowers open or close they do so under the influence of light or air, or of some other cause not requiring a brain or nervous system to produce it. In the "sensitive plant," as it is called, the leaves collapse under the slightest touch, but this is merely a power of contraction not dependent on a nervous system.

Nutrition of Animals and Vegetables.—So far as nutrition is concerned in animals and vegetables, there is observable a marked difference. The source of supply of food to the vegetable is the earth which surrounds it and in which it is firmly fixed. The food thus offered it does not require preparation ; it is ready for use, and it is at once absorbed by the vegetable. Whatever the animal needs to sustain it must be taken into the stomach and digested, as it is called, or prepared for absorption for its nourishment. In both animals and vegetables absorption of the food is necessary before it can be of any service in maintaining life; but the higher classes of animals also have the power of seizing upon their food, of moving to obtain it, and, as already stated, a will to govern them in their movements.

Chemical Composition.—As a rule, vegetables are mainly composed of the three chemical substances, oxygen, hydrogen, and carbon; animals have all these, and nitrogen in addition, a fact which may be shown as follows:

$$\text{Vegetable......} \left\{ \begin{array}{l} \text{Oxygen,} \\ \text{Hydrogen,} \\ \text{Carbon,} \\ \text{Nitrogen,} \end{array} \right\} \text{......Animal.}$$

This is not wholly true of either animals or vegetables, for the green parts of vegetables are made up of the four elements, and the fatty parts of animals of the three elements only. This explains why in cases of excess of fat in individuals animal food is given as a means to reduce it, rather than starchy matter or sugar, which are made up of the three elements, for these last would only increase the evil which it is attempted to remedy.

In addition to the four elements just mentioned, many others are found in the animal tissues. Some of the principal of these may be conveniently arranged in a table, which gives at a glance the particular parts of the body in which they have been detected. It will be observed that the most important of these are found in all the tissues and liquids, while others have a special location:

NAME.	WHERE FOUND.
Hydrogen	In every tissue and fluid.
Carbon	In every tissue and fluid.
Nitrogen	In many of the tissues; in solution in fluids.
Oxygen	In all the tissues; in solution in fluids.

NAME.	WHERE FOUND.

Sulphur.........Albuminous substances; blood; secretions; serum of the tissues.*

Phosphorus....Blood; nervous matter; bone; teeth; fluids.

Chlorine.........In every tissue and fluid.

Sodium..........Blood; all the secretions; serum of the tissues.

Potassium......Muscles; red blood-corpuscles; nervous matter; secretions

Calcium.........Bones and teeth; fluids.

Magnesium....Bones and teeth; fluids.

Lithium.........Muscles; blood; milk.

Iron..............Coloring-matter of the blood; bile; chyle; lymph; sweat, etc.

These simple chemical elements are combined to form other bodies or materials of which the various animal textures are composed. Water is one of the most important of these. The chemical composition of a living being varies in quality and quantity at different periods of its growth. The seed, for example, differs chemically from the plant; the egg from the grown animal. The quantity of water gradually diminishes also as life advances. It usually forms about two-thirds of the weight of the body. In a person weighing one hundred and fifty pounds there are about one hundred pounds of water. It is distributed through the body in the most important organs as follows, in each 1000 parts:

TISSUES OR ORGANS.	WATER.	SOLID PARTS.
Enamel of the teeth....................................	2	998
Bone..	220	780
Fat...	299	701
Cartilage..	550	450
Liver...	693	307
Spinal cord..	697	303
Skin..	720	280
Brain...	750	250

* By *secretions* are usually meant fluids separated from the blood, as the sweat, the tears, etc.; by the *serum* of the tissues, the thin watery portions which bathe their surfaces like a moisture.

2 *   B

TISSUES OR ORGANS.	WATER.	SOLID PARTS.
Muscles	757	243
Blood	791	209
Bile	864	136
Milk	891	109
Chyle	928	72
Gastric juice	973	27
Tears	982	18
Saliva	995	5

There are also in various parts of the body, as in the gastric juice, bones, muscles, blood, bile, etc., *acids*, such as muriatic, sulphuric, phosphoric, lactic, carbonic, and other acids; *alkaline substances*, such as soda, potassa, magnesia, and lime; *salts* of various kinds, such as phosphate of lime, chloride of sodium, or common salt, etc.; *sugars*, such as sugar of milk; *fats*, as in the oily part of milk; *coloring-matters*, as in the blood and bile; and *albu'minous matters*, which we shall hereafter find to be present in the blood, milk, brain, muscles, and other important solids and liquids. The white of egg is the best example of albumen, and like albuminous substances in the various parts of the body is coagulable by heat, a fact familiar to all in the appearance of the egg when boiled. *Vegetable albu'men* is the name given to it when found in the juices of plants. In milk it is called *ca'sein;* while in peas, beans, etc., and in the grains of similar plants, it is called *legu'min*. The albuminous body formed during the coagulation of lymph and blood is called *fi'brin*. In the red globules or corpuscles of the blood there is an albuminous body called *glob'ulin;* and when the albuminous matters in the stomach are acted upon by the gastric juice, a substance called *peptone* results.

In addition to all these and other substances in the bodies of animals, certain airs or *gases* are found. The

air that passes into the lungs when we breathe is mainly
composed of oxygen and nitrogen. From food and drink
taken into the stomach we derive such gaseous matters as
are found in that organ or in the intestines. The principal
gas given off from the lungs is carbonic acid, which comes
from the breaking down of the tissues, and has its outlet
from the system chiefly at that point.

Cells.—All the tissues and organs of the body originate
from a minute form called a *cell*, which divides into other
cells, and these, by uniting together, are developed into
tissues. *Organs* are made up by a combination of tissues.
Several organs grouped together form a *system* or *apparatus*. In the apparatus of digestion, for example, there
are several organs, such as the stomach, intestines, etc.,
which are made up of different kinds of tissues.

The cells are small vesicles, only visible under the microscope, composed of a very thin membrane called a *cell-wall*,
and contain a semi-fluid matter, in which is frequently
imbedded a minute oval body called the *nu'cleus*, itself
containing sometimes an even smaller body, the *nucle'olus*.
So that we may correctly say that the nucleolus is the very
lowest form in which a tissue is capable of being expressed.
To give an idea of the size of the cells, it may be stated
that some of the smallest of them are so very minute that
12,000 of them placed together would only occupy the
space of an inch, a measurement which the mind scarcely
realizes. Fluids pass through the walls of these cells
so readily that the latter may swell when brought into
contact with thin fluids, while they may shrivel up if
their own contents pass out into a thicker fluid around
them.

Where Nutrition is Effected.—It is in these cells that
the nutrition of the body is effected. They are found in

the vegetable as well as in the animal. The cell receives
certain materials from the surrounding medium, and converts them into its own substance or makes use of them
for nutritive purposes. Certain cells have a power of
selection by which special kinds of material are formed.
Some cells, for instance, form coloring matter, some fatty
matter, etc. They make this selection or choice as perfectly as if they had a mind of their own to control them.
The cells are endowed with special qualities by which they .
perform certain actions when stimulated. Thus a cell of a
muscle contracts when it is stimulated; the cell of a gland
pours out, or secretes, as it is called, a fluid,—that of the
perspiration or the tears, for example; and a nerve-cell
takes on itself an action which results in sensation or intellectual power or exercise of the will. Each cell has a life
of its own separate from all those which surround it; but
the life of the whole body is made up of the lives of all
the cells which compose it. Some cells die after a few
hours; others may exist for years. The cells on the surface of the body, for example, are being constantly removed and renewed, while those of cartilages, structures
which have not much vitality, may endure for a long
time. The importance of cell-agency in the nutrition of
the various organs must not be under-estimated because
these bodies are microscopic.

Some of the important cells of the body, as the nerve-
cells, blood-cells, or corpuscles, etc., will be alluded to
hereafter when we come to describe the nervous system,
the circulation of the blood, etc. The solid fat of the
body is contained in separate cells with delicate walls,
called *fat-cells* or adipose-cells. They are globular in
shape, but become many-sided when pressed together, and,
although usually larger, are often as small as the $\frac{1}{3000}$th

of an inch in diameter. Some cells, known as *pigment-cells*, contain a coloring-matter in their interior, and are found in certain parts of the eye and on the skin. They are concerned in producing the different colors of the races of mankind, and in the eye they have the effect of assisting in absorbing the rays of light, like the dark parts of the interior of a telescope.

Divisions of the Human Body.—When we study the human body in its various parts and structures, we find it made up of a number and variety of organs. The framework is the *skeleton* (Fig. 1), which is composed of more than two hundred bones. The skeleton has at its upper part the *skull* (Fig. 2), itself a collection of bones of various sizes, fitted accurately together, containing in its interior the different portions of the brain. Continuing downward from the skull, which is carefully poised upon it, is the spinal or *vertebral column*, or *spine*, which is a series of bones, in the interior of which is the spinal marrow. At the lower part of the spinal column, on each side, are large expanded bones, called the peivic bones or *pelvis*, from their resemblance to a basin, as their Greek derivation implies. The sides of the pelvis form the hips. The framework of the upper and lower limbs is also bony, being composed of bones of various sizes adapted to the useful movements of those parts. These bones, which make up the skeleton, are covered with *muscles*, which give motion to the various portions of the body. Embraced within these various walls formed by bones and muscles are the important cavities of the body—the *head, chest,* and *abdo'men*—in which the organs concerned in sensation, respiration, circulation, digestion, etc., are placed. In the skull is the brain, and in the face, which is the front portion of the skull, are the organs of sight, smell,

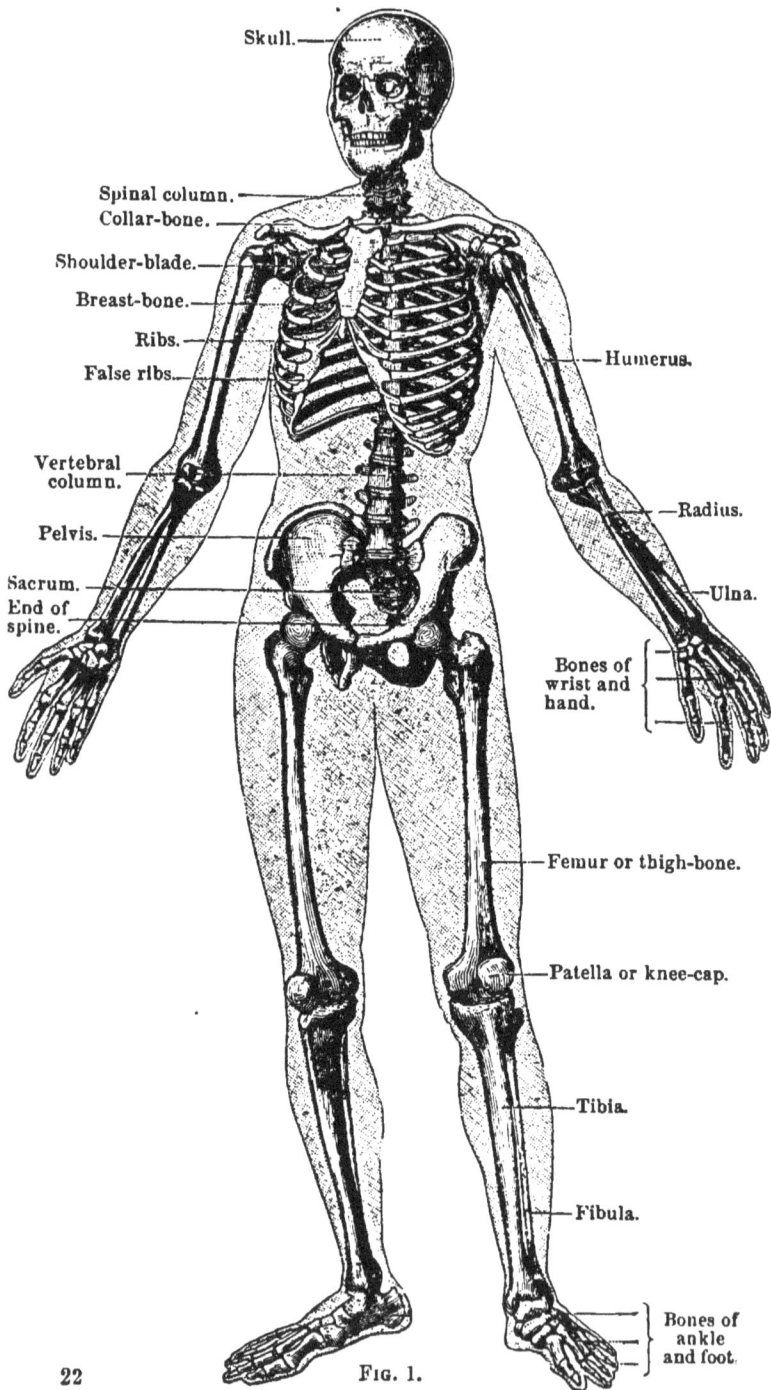

Skull.

Spinal column.
Collar-bone.

Shoulder-blade.

Breast-bone.

Ribs.

False ribs.

Humerus.

Vertebral
column.

Pelvis.

Radius.

Sacrum.
End of
spine.

Ulna.

Bones of
wrist and
hand.

Femur or thigh-bone.

Patella or knee-cap.

Tibia.

Fibula.

Bones of
ankle
and foot.

22

Fig. 1.

taste, and hearing. In the chest, or thorax, as it is called, are the heart and lungs; in the abdomen are the digestive organs, in which the food is acted upon and absorbed for the use of the system.

In addition to the bones and muscles and the important organs just referred to, there are numerous *bloodvessels,*

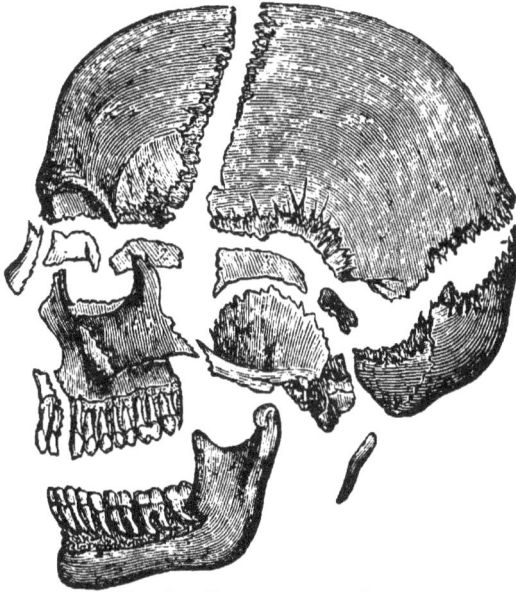

FIG. 2.—BONES OF THE SKULL.

conveying blood from the heart to all parts of the body, and back again to the heart, and called *arteries* and *veins;* and also a set of vessels in every region, called *lymphat'ics,* which convey a thin fluid, called *lymph,* to be mixed with the blood. Delicate cords, called *nerves,* greatly varying in size, pass everywhere throughout the system, like tele-graph-wires, keeping up a communication between the brain, spinal cord, and every portion of the body. Cover-ing the whole mass of muscles, bones, vessels, and nerves is the *skin,* the sensitive envelope which protects and shields the whole body from external irritants.

QUESTIONS.

Into what three classes have bodies been divided?
In what respect do two of these classes resemble one another?
What are organized bodies? Inorganic?
To which class do animals belong? Vegetables? Minerals?
How do you define Physiology?
What is its derivation?
What is Vegetable Physiology?
When the word Physiology is used alone, to what does it apply?
What is Human Physiology?
What is Comparative Physiology?
What does the study of Physiology embrace?
What is Anatomy?
What relation does Anatomy bear to Physiology?
What information does the anatomy of the brain give us?
What is its chemical constitution?
What do we learn of physiology from the anatomy of the eye or tongue?
What is Comparative Anatomy?
What assistance does this give in learning physiology? Take the stomach as an example.
What may we learn from disease or the absence of an organ? Of the brain, for example?
What examples of physical apparatus have we in the body?
What process is chiefly chemical?
What part does chemistry take in it?
In what respect is respiration a chemical process?
What action has a nervous system in such processes?
What is the healthy action of the heart dependent upon?
How do we explain this continuous action of the heart?
How do inorganic bodies differ from animals in regard to life, size, shape, etc.?
Why is change necessary to the life of organized bodies?
What evidences of sympathy or dependence of different parts have we in animals and vegetables?
How does this differ from the mineral or inorganic substance?
What are the various organs called in the vegetable? In the animal?
How are tissues made up in the vegetable or animal?
What is meant by the different ages in animal life?

Under what circumstances do inorganic bodies undergo change?

When does death or the cessation of life occur in organized bodies?

What is said of the death of inorganic bodies?

When are organs said to be healthy?

What is disease?

What does hygiene include?

What function or process is common to animals and vegetables alike?

What two functions have animals that vegetables do not possess?

Are solids or fluids in excess in the vegetable? In the animal?

What is the elementary tissue of the vegetable?

How does this differ from the animal?

Why is a nervous system not necessary to the vegetable?

What peculiar exceptions are noted in the lowest forms of animal and vegetable life?

How do flowers open and close? The "sensitive plant"?

What is the source of supply of food to the vegetable?

What difference is noted in regard to the animal?

After taking food, in what process of nutrition do animals and vegetables resemble one another?

What additional facilities for nutrition have the higher classes of animals?

Of what three chemical substances are vegetables mainly composed?

What additional substance is found chiefly in animal tissues?

What parts of the vegetable are made up of the four elements?

What part of the animal is composed of the three elements?

What kind of food should be given to those suffering from excess of fat?

Why should not starchy or fatty matters be given in such cases?

State from the table what other chemical elements are found in the body.

Which of those named are found in all the tissues and fluids?

Select from the table those other elements which are found in the blood. In the nervous system. In muscles. In the bones and teeth.

What special element is found in the coloring-matter of the blood?

What do we mean by secretions? By the serum of the tissues?

What proportion of the weight of the animal body is fluid?

How much water is there in a person weighing one hundred and fifty pounds?

What parts of the body are shown by the table to contain more solids than fluids?

Mention a few important fluids that are very largely made up of water.

3

What acids are found in various parts of the body? What alkaline substances? What salts? What other chemical substances?

In what parts of the body are albuminous matters present?

What effect has heat upon albumen?

What name is given to this substance as found in the juices of plants?

In milk and beans what names are given to it?

What is fibrine?

What is the albuminous body in the coloring-matter of the blood called? In the stomach as the effect of digestion?

How are oxygen and nitrogen taken into the system?

What is the principal gas given off through the lungs?

From what minute form are tissues and organs developed?

How do organs differ from tissues?

What is a system or apparatus?

What is included in the apparatus of digestion?

What are cells composed of?

What is a nucleus? A nucleolus?

What is the lowest expression of a tissue?

What is the size of the cells?

Where and how does nutrition take place?

What power of selection do cells possess?

What action takes place when a muscular cell is stimulated? A gland-cell? A nerve-cell?

How long is the life of a cell?

What is the arrangement of the fat-cells?

What are pigment-cells, and where do they exist?

What effect have they on the skin or the eye?

What is the bony framework of the body called? How many bones compose it?

What is the skull, and what does it contain?

What is the vertebral column? What does it contain?

What are the pelvic bones?

How are the bones covered?

What organs are contained in the front part of the skull? In the chest? In the abdomen?

What organs are concerned in conveying blood to all parts of the body?

What are the lymphatics?

What organs maintain the communication between different parts of the body?

What general office does the skin perform?

THE BONES, JOINTS, AND MUSCLES.

The Bones.—The general shape and solid form of the body depend upon the bones. These act not only as a framework for support, but also as a protection to the delicate organs which they enclose. Some very important organs of the senses are thus protected, as the eye and ear, within the bony structures surrounding them. The brain itself is completely sheltered from ordinary injuries by the skull, which, as a bony, air-tight case, covers it at every point. The heart and lungs are also similarly protected by the bony and muscular walls of the chest. The bones are also useful as offering surfaces on which the muscles are inserted, for on the bones and muscles thus acting on each other depend the movements of the body. The important part of the nervous system included in the spinal cord is protected by the spine, or spinal column, which is a long collection of bones extending the whole length of the back.

The bones are of a great variety of shapes: some of them, as those of the hands, are quite small; others, as of the arms and legs, large and capable of numerous movements. Bone consists of an earthy and an animal part resembling gelatin. The earthy part is chiefly phosphate of lime, and when bone is acted on by heat and acids this is destroyed, and the bone becomes so flexible that it can be tied into a knot. This is like restoring it to the condition it was in when it was first formed, for it was gelat-

27

inous before it became encrusted with earthy matter. The animal part of the bone imparts to it elasticity and renders it tough, while the earthy portion gives it rigidity. The earthy matter is in excess in old persons, hence the bones are more liable to break at that period of life.

Cartilage.—Bony matter exists at first in a kind of pre-paratory condition called *cartilage*, the cells of which become replaced by bony cells, which give to bones their particular consistence. Some portions of the body, as parts of the ribs, remain during the whole period of life in this state of cartilage. Where the bones come together to form a joint, as in the elbow or knee, the surfaces are in this very condition. In the skull, made up of a number of bones of different shapes and sizes, no such arrangement is necessary. Almost all the bones here fit into one another by their opposite edges being bevelled and grooved, so that they can be accurately adjusted (Fig. 1. The bones are here shown as if widely separated, so that their edges may be more plainly seen). The bones of the skull proper are not movable, the joints being more like seams, and hence called *sutures*. The lower jaw is the only part that is movable, but it is connected with the skull in a different manner from that seen in the other bones.

Division of Bones.—Bones are generally divided into long, short, and flat bones. The thigh, for instance, is a long bone; those of the skull are flat. This is a good division so far as the bones of man are concerned, but not of other animals as compared with man, for the same bones that we might call long or flat in the one might be of very different shape in the other. Some small bones are called long, because their structure internally is like that of long bones.

Structure of Bones.—If we examine the interior of a

long bone by sawing through it lengthwise, we are able to understand how both strength and lightness are imparted to it. It will be seen that it is not solid throughout, as might appear from an external view of it; but it is divided into two distinct portions (Fig. 3)—a hard part, which is also porous and honeycomb-like at the extremities of the bones, and a spongy portion, containing in its centre a substance called the *marrow*. When we examine the structure of bone under the microscope, we notice an appearance similar to that presented in Fig. 4, the bony mass being hollowed out by an immense number of small canals, called the *canals of Havers*, after the anatomist who first described them.

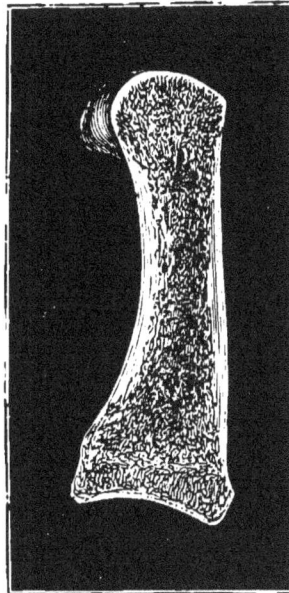

FIG. 3.—INTERIOR OF A BONE.

Through these canals pass the bloodvessels which nourish the bone and the marrow contained in it. The bony canal is a sort of star-shaped cell, as seen in the section represented (Fig. 4).

The Spinal Column.—This, generally known as the *backbone*, is sometimes also called the *ver'tebral column*, each piece forming it being called a *ver'tebra*. (The general arrangement will be seen by reference to Fig. 5.) The various *spines*, or projections, give

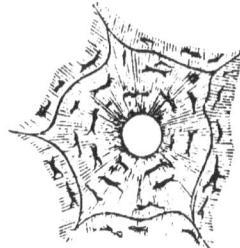

FIG. 4.—SECTION OF BONE.

it the name of the spinal column, by which it is generally known. These are readily felt by the hand along the

2 *

middle of the back.　There are twenty-six bones in the chain, smaller at the top of the column, and larger as it descends.　It is the back-bone or groundwork which acts as a basis of support to the chest and abdomen (Fig. 6).　A canal, called the *vertebral* or *spinal canal*, passes through the interior of the spinal column, and contains the spinal cord, which is one of the great nervous centres.

Spines, or spinal processes.

Vertebræ of the neck.

Vertebræ of the back.

Vertebræ of the loins.

Between the bones composing the spine are layers of elastic tissue, which give to the whole column a slight degree of flexibility and elasticity without imparting much power of movement to it.　The useful effect of this elastic tissue or cartilage placed between the harder texture of the bones is also to diminish the force of blows and injuries inflicted upon the spine.　Were this not present, the force of the slightest falls or jumps would be felt by the delicate brain as transmitted along the solid bony mass

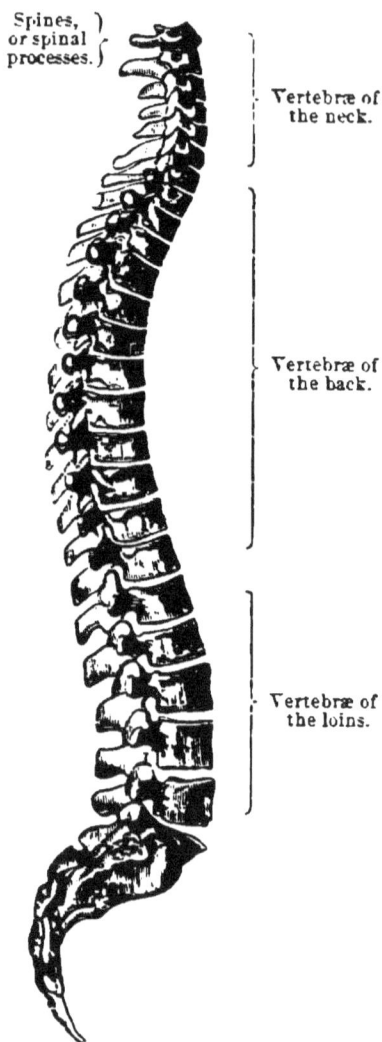

FIG. 5.—SPINAL COLUMN.

of the spinal column.　The effect may be illustrated by a simple and well-known experiment.　If a number of solid

ivory balls be suspended by strings parallel with one another, and the last one of the series be raised and allowed
to fall against its neighbor, the shock will be communicated
through all the balls, and the first one in the row will fly
off at a tangent. If,
however, one, or per-
haps two, porous balls
be placed in the row,
and the last ivory ball
be again brought into
contact as before, the
force of the blow will
be so greatly broken
that the first ball will
probably remain sta-
tionary. The elastic
cartilages between the
spinal bones yield so
much to pressure dur-
ing the day that a per-
son is actually shorter
in stature at night
than he is in the
morning after a night's
rest has restored them
to their natural con-
dition.

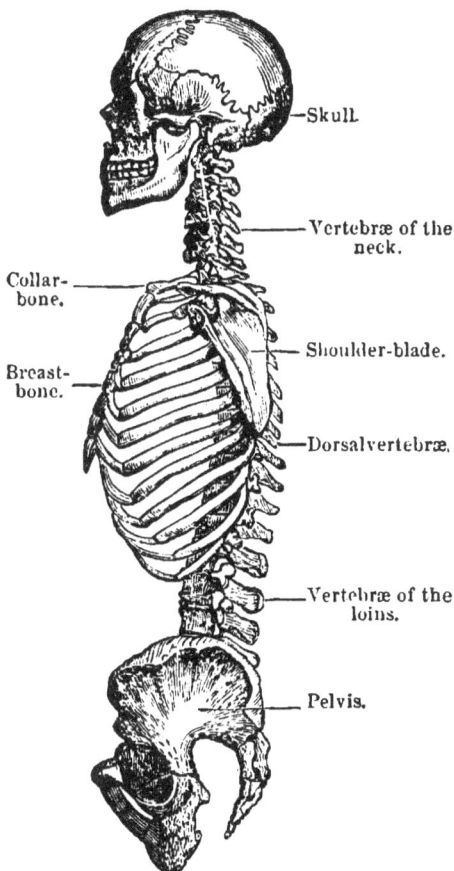

The relative posi-
tions of the spinal column, the skull, and the chest are
seen in Fig. 6. The spine is firmly planted between two
large, irregularly-shaped bones, called the *pelvic bones*
(from a Greek word meaning "a basin") or *nameless bones*,
because they have no special resemblance to any familiar

FIG. 6.—SKULL, SPINE, ETC.

Skull

Vertebræ of the
neck.

Collar-
bone.

Shoulder-blade.

Breast-
bone.

Dorsal vertebræ.

Vertebræ of the
loins.

Pelvis.

object.   Into the socket or cavity at the side of each of
these bones is inserted the thigh-bone.   This mode of
insertion of the spine into these bones has been compared
to the manner in which the mast of a ship is inserted into
the keelson. and the strong ligaments which connect the
bones together have been likened to the shrouds which
bind the mast to the sides of the vessel.   The head is connected to the spine so as to admit of freedom of motion in
various directions.   It was necessary to provide a mechanism that would allow the head to rotate or turn from side
to side or to nod, as required.   The two upper bones of
the spinal column are differently arranged from the other
bones of the series to accomplish this very result.   The
upper bone is called the *atlas*, for in mythology it was
Atlas who bore the globe upon his shoulders; the second
is called the *axis*. because upon it the head turns.   The
bony cage (Fig. 6) attached to the spinal column is the
thorax, which is made up of ribs and the breast-bone, with
the spine at its back part.   Its arrangement and uses
belong particularly to the consideration of Respiration, or
the process of breathing, for in the thorax or chest are
contained the lungs.

  The Limbs (Fig. 1).—These are known as upper and
lower limbs.   The *upper limbs* consist of one arm-bone, two
bones of the forearm—the ra'dius and ulna—and twenty-
seven bones of the wrist and hand.   The *lower limbs* are
composed of a nearly similar arrangement of bones,—one
of the thigh, two of the leg—the tib'ia and fib'ula—and
twenty-six bones of the ankle and foot.   The arm-bone,
generally known as the hu'merus, moves in the socket of
the shoulder-blade. a bone which lies on the upper and
back part of the thorax (Fig. 6).   The femur, or thigh-
bone, as already stated, has its motion in the cavity or

socket of the hip-bone. These bones are all attached to one another by bands or *ligaments*.

The *lower animals* differ from man in the number of bones, but chiefly in their general arrangement for loco-

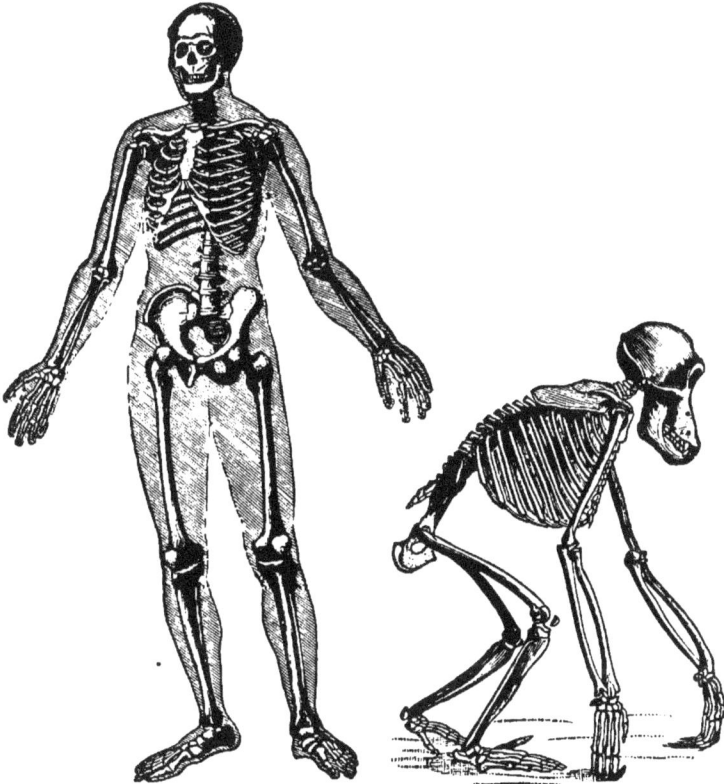

FIG. 7. SKELETONS OF MAN AND CHIMPANZEE.

motion or movement from place to place. The stature of the chimpanzee, for example (Fig. 7), is not erect like that of man, and its skeleton is made to conform to its general movements, the bones of its limbs being longer in proportion than those of man, and aiding in its stooping attitude.

C

We cannot dwell upon these differences as we descend in the scale of animal life without entering too minutely into the subject of Comparative Anatomy.

The Perios'teum.—The surfaces of bones are covered by a fibrous membrane, through which bloodvessels and

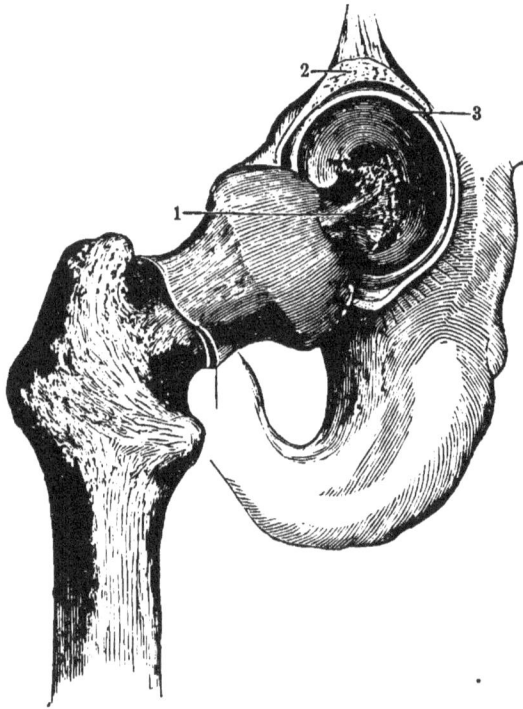

FIG. 8.—HIP-JOINT.*
1, 2, 3, ligaments.

nerves pass for the nutrition of the bone, called the periosteum, strictly from its derivation from two Greek words, meaning "over or about the bone."

The Joints.—A passing allusion has already been made

---

* The bone in Fig. 8 is represented as unnaturally moved from its socket, so as to show the cavity. In Fig. 9 the ligaments of the elbow appear as if separated, for the same reason.

to some of the modes in which bones are separated from one another. Strictly speaking, we might say that the bones are never thus separated absolutely, for they are attached to each other at every part of the body, although in very different ways. Some of the joints—or *articulations* as they are called—are movable; others are immovable. Those that are movable require strong ligaments or bands to keep them sufficiently in their place to prevent dislocation. *Lig'aments* are white fibrous cords, which yield so little when the bone moves in the joint that sometimes when great force is applied the bone will break and the ligament sustain no injury. When violently stretched or twisted, however, what is generally called a "sprain" or "strain" results. In some articulations the surfaces are covered with a fibrous and elastic substance called *car-tilage*, which is able to resist the application of great force

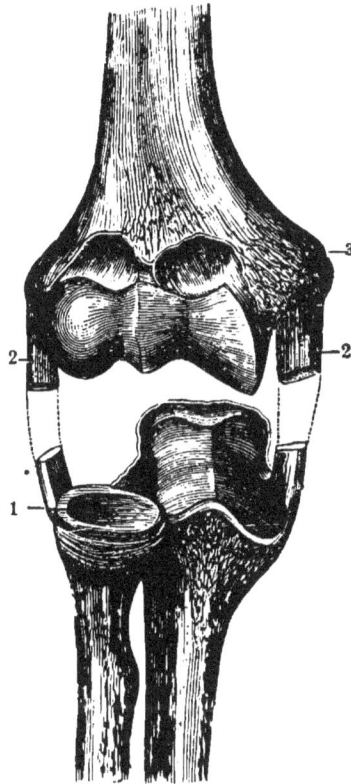

FIG. 9.—ELBOW-JOINT.

1, bones of forearm.; 2, ligament; 3, bone of arm.

or shocks. This cartilage is bathed by a fluid called the *syno'vial fluid*, on account of its resembling white of egg, which allows the surfaces to move smoothly over one another. The machinery is thus lubricated, as it is familiarly called in the phraseology of the workshop when oil is applied to the joints and working-parts of the steam-engine. Strength and ease of motion are thus imparted to the joint by the ligaments, cartilages, and synovial fluid combined. The greatest freedom of motion is seen in the shoulder-joint, which is a familiar illustration of a ball-and-socket joint; another example of which is seen in the hip-joint (Fig. 8). The elbow, knee, and wrist have a hinge-like movement familiar to all. The mode in which the different bones fit into one another, and the arrangement of the cartilages, cavities, etc., are well shown in Fig. 9.

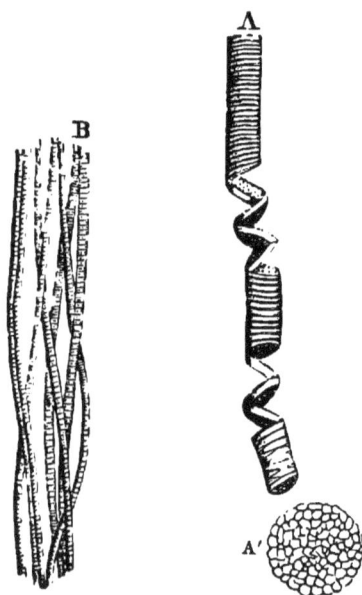

FIG. 10.—FIBRILS OF MUSCLE.

A, B, muscular fibrils; A', cross-section of one of the discs composing them.

The Muscles.—The erect position of the body is maintained chiefly by the action of the muscles in connection with the bones. These are a series of organs, familiarly known as the *flesh*, which abound in every part of the body, giving it shape and directing its movements. The muscles are about four hundred in number. *Meat*, as it is familiarly called, is the muscular portion of the animal body which is used for food.

Muscular Fibres.—When a muscle is examined under the microscope it is found to be made up of a large number of small fibres, or *fi'brils*,* as

---

* The general arrangement of the fibrils when deprived of the membrane covering them is seen in Fig. 10, A, one of the rings composing it being here represented, still further enlarged.

they are called, too minute to be seen by the naked eye, and parallel to one another. These fibrils (Fig. 10, B) are the contractile portions of the muscle. A large number of them are united into bundles, which are placed parallel with other bundles, from which they are separated by a

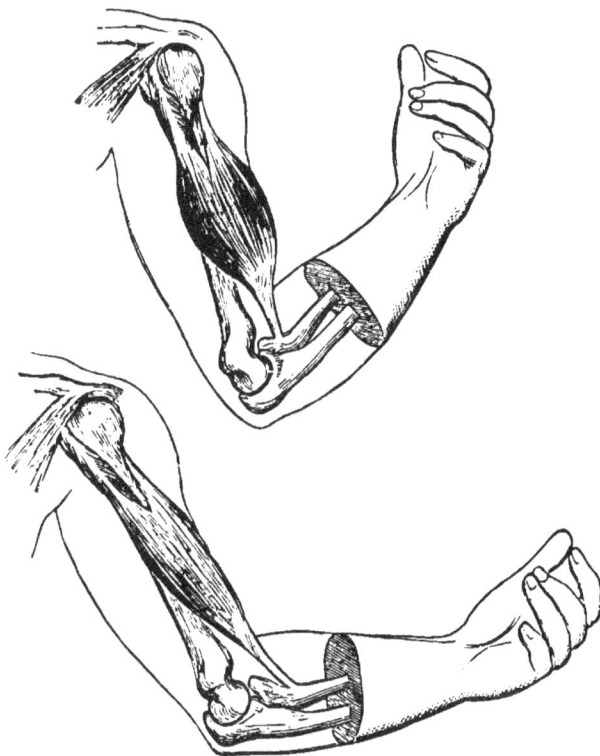

FIGS. 11, 12.—ACTIONS OF A MUSCLE ILLUSTRATED.

loose tissue called are'olar or cell'ular tissue. These bundles combine with other bundles, and so on. The muscles are usually placed between or in contact with bones, sometimes by both extremities of the muscle, or, as in the case of the eye, by one end only. The action of muscles, wherever they exist, is motion, in some form. Sometimes the motion is voluntary; at other times, as in the heart, which is a

4

muscular organ, it is involuntary, or beyond the control of the will. As a general rule, the voluntary muscles are striped by parallel lines, and hence the term *striped* or *stri'ated* muscles applied to them. The heart is an exception, being striped and yet involuntary. All muscular movement that occurs during sleep is carried on by involuntary action of the muscles. Those which are concerned in breathing, by raising and depressing the ribs or moving the muscles of the abdomen, are partly voluntary and partly involuntary, so far as the nervous system controlling them is concerned.

Tendons. — One of the best examples in the human body of a muscle producing voluntary motion is that generally known as the biceps of the arm, so called because it has two heads or origins (Figs. 11, 12). The effect of its contraction is shown to be the bending of the forearm at the elbow in a direction toward the shoulder.

FIG. 13.—TENDO ACHILLES.

The natural appearance of the muscle when in a state of repose is also exhibited. The fibres of muscles terminate (as seen in Figs. 11, 12,) in a fibrous non-contractile tissue, to which the name *tendon* has been applied. By this tendon it is firmly inserted into the bone, and all direct action of the muscle on the bone in moving the limb is exerted through it. Some of the tendons just beneath the surface of the skin can be felt at different points, such as the back part of the knee and at the elbow. The most powerful tendon in the body is that which is inserted into the heel, and is controlled by the muscular mass on the back of the leg (Fig. 13). It is known as the tendo Achilles, from the fact that this was, in mythological story, the only vulnerable part of the body of Achilles. Another fibrous arrangement connected with the muscles is called an *aponeuro'sis* (because it resembles nervous tissue, according to the notion of the ancients). It is a white, shining membrane, very resisting, and differing only from the tendons in its flat form. It is sometimes continuous with the muscular fibres, and at other times forms a covering for them, surrounding them so as to prevent their displacement. In the extremities it not only invests the whole limb, but gives off portions to pass in between the various muscles (Fig. 14).

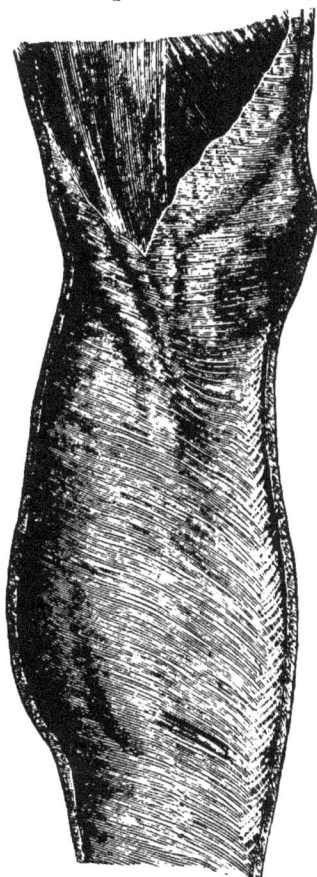

Contractility.—The characteristic property of the muscles is their power of contraction, or *contractility*, as it is called. The result of this contraction is to diminish the muscle in length,

FIG. 14.—APONEUROSIS OF THE LEG.

while at the same time it becomes broader or swollen from side to side. The shape of the muscle as a fleshy mass therefore becomes considerably changed (Fig. 11). The hardness which is felt when the hand is placed over the muscles of the arm when the forearm is flexed or bent at an angle to the arm is caused by the forcible tension of the muscular fibres produced by their contraction. The changes, therefore, which a muscle undergoes in contracting are a shortening of its length, an enlargement in its thickness, and a greater degree of firmness and tension. After contraction, which lasts for only a short time, and results in fatigue if long continued, the muscle returns to its previous condition of repose. The parts on which it had acted while in a state of contraction also resume the position they had assumed when influenced by that muscle or series of muscles, or assume some new position; but this effect is not produced by the muscles which had already acted upon them, but by another set, which, from their acting in opposition to them, are said to be *antagonistic muscles.*

Muscles that bend the joints are called *flexors,* and those which restore the part to its straight condition are called *exten'sors,* and these two sets of muscles are excellent illustrations of the antagonism already referred to. The muscles of the arm and the leg which bend or straighten those parts are flexors and extensors respectively. Although called antagonists, they do not oppose each other or counteract their individual movements while either set is on duty; each muscle or set of muscles—whether flexor or extensor—waits until its opponent has performed the work assigned to it. Should both sets of muscles be called into play at once, the result would be complete rigidity and absence of motion in the limb. Fatigue soon results, however, from such a forced condition of the muscular apparatus. Rest is absolutely necessary for the muscles, even those which, being beyond the control of the individual, have been called the involuntary muscles. The heart, for example, which is supposed by many to be always at work and never to rest, has really a brief period of repose between its pulsations or beats. Muscles require such a rest to enable them to recover from their fatigue and to acquire new strength. The fatigue of keeping a limb, such as the arm or the leg, in an outstretched position for any length of time is a familiar illustration in point, whereas if the part was properly exercised the alternate contraction and relaxation would preserve the tone and vigor of the muscles.

Movements.—The movements of the body in walking, running, leaping, etc. are examples of the influence and utility of the muscles.

Walking is apparently a simple matter of every-day experience, but it is a series of complicated movements of the flexor and extensor muscles of the legs. It is motion on a fixed surface, the centre of gravity of the body being moved alternately by one of the extremities and supported by the other without absolutely leaving the ground. In walking the limb is directed forward, and the weight of the trunk is supported by the hip-joint—that is to say, by the head of the thigh-bone. This weight is transmitted to the ground by the principal bone of the leg and by the bones of the foot. Then the opposite lower limb advances in its turn, and supports alone the weight of the body, while the limb that is behind is in repose. For a single moment, but a very short one, in these successive steps, the body rests upon both legs at once. The essential point in the process of walking is the fact that the heads of the thigh-bones form fixed points, on which the pelvis turns alternately as upon a pivot, describing arcs of circles, which are greater in proportion to the size of the steps.

In *running*, the foot that remained behind is detached from the ground before the other reaches it, and there is not a single instant in which the body rests upon both feet

4 *

at once. At some moments, indeed, both feet are entirely
raised from the ground. Running is really a succession
of low leaps performed by each leg alternately. It differs
from walking in the fact that the body is projected forward at each step, and the hindmost foot is raised before
the foremost foot touches the ground. In running, the
body is inclined forward, so that the centre of gravity
may be properly placed to receive an impulse in that
direction from the hindmost leg; and the foremost leg is
advanced quickly to keep the vertical line within the base
of support, and thus prevent the body from falling.

In *leaping*, the whole body is raised from the ground,
and the limbs are suddenly extended after violent flexion.
Swimming resembles leaping, but there is no fixed surface
from which the muscular effort is made. Many other muscular movements are possible, such as are called into play
in climbing, carrying a load, etc., but it is not necessary to
enter more fully into an explanation of them.

Effect of Exercise.—As the bones give general outline to
the body, so do the muscles fill up all the details, giving
the individual his characteristic features, shape, etc. The
development of the muscular system is dependent upon the
amount of exercise to which it is subjected. We therefore
find such occupations as call the muscles into active play
producing in them greatly increased development; the
blacksmith, for example, having the muscles of his arm,
especially the biceps previously referred to, largely increased in size and power. The amount of development
varies also with the age and with the sex, being generally
slight in children and in females. All development, however, must be gradual, and not forced or violent. All the
attitudes practised in gymnastics are intended to have the
effect of strengthening the muscles of the arms and legs

and chest, but great care must be taken not to exceed moderation in the exercise, as the muscles may be developed at the expense of the powers of the individual. A familiar effect of the gradual exercise of the muscles under proper training is seen in the preparation of horses for racing. They are brought very slowly into proper condition, and at first they exercise without any weight upon their backs, and afterward, but gradually, with a weight that is increased by degrees up to the proper standard.

During waking hours all the muscles are never in repose at one time. While standing, the extensor muscles of the limbs and the muscles which support the head and spine are in a state of tension. When we sit down without having any support to the back, the muscles which support the spinal column and the head must be in a state of tension, and if the back be supported, then the muscles of the head must be in a like condition. During sleep the power of the will over the muscles is subdued and the muscles become relaxed, and each part falls or drops, as the head upon the chest, according to the laws of gravity. It is easy to understand why a person becomes easily fatigued from such continuous tension. Repose or rest of the muscles concerned is absolutely necessary.

Names of Muscles.—All the muscles of the body have received names, but these are important only to the anatomist, one who makes a special study of all the different minute parts of the body. We need only know them in classes by the work they have to do. We have already spoken of *flexors* and *extensors;* the flexor muscles of the leg, for example, passing from the back of the thigh to the back of the leg, and the extensor muscles from the front of the thigh to the front of the leg. Then we have muscles which draw the arm or the leg away from the body, and

these are called *abduct'ors,* and others which move them
toward the body, and these are called *adduct'ors.* If a
part is lowered by a muscle, the latter is called a *depress'-
or;* if it compresses a part, it is called a *compress'or;* if it
dilates, a *dila'tor.* It has been already shown in illustrating the bones that they are arranged in pairs (Fig. 1); that
is, that both sides of the body, the right and left, have, as
a general rule, the same bones on both sides of a median
line. The same remark is true of the muscles, which are
alike on both sides of the body, except when the muscle is
single and divided into two parts by the line drawn through
the middle of the body.

Muscles of Expression.—Besides giving motion to the
limbs and bones in various parts, the muscles have other
important duties to perform. The bones of the face, for
instance, are not at all movable upon one another, except
in the case of the lower jaw, yet the whole expression of
the face in its various emotions, as joy, anger, grief, etc., is
dependent upon the play of the muscles connected with the
soft parts of that region (Fig. 15). In some parts of the
body a great deal of work is performed by the contraction
of muscular fibres that are almost, if not wholly, microscopic, as in the smaller air-tubes, in which the action of
the muscles assists in the process of breathing. In the face,
the muscles connected with the eyelids, nose, eyebrows, lips,
cheeks, tongue, etc. all assist in the expression of the emotions, and combined they give to each individual his identity
and peculiar features. The motions of the face and neck
are controlled by about seventy pairs of muscles in different
layers.

When muscles are allowed to remain unused for any
length of time they lose their firmness and are lessened in
size, becoming softened and weakened in power. Seden-

tary occupations and indolence have the same effect, and the whole system sympathizes, so that the general health suffers. Exercise in the open air is absolutely necessary each day, and should not be crowded into one day, to the

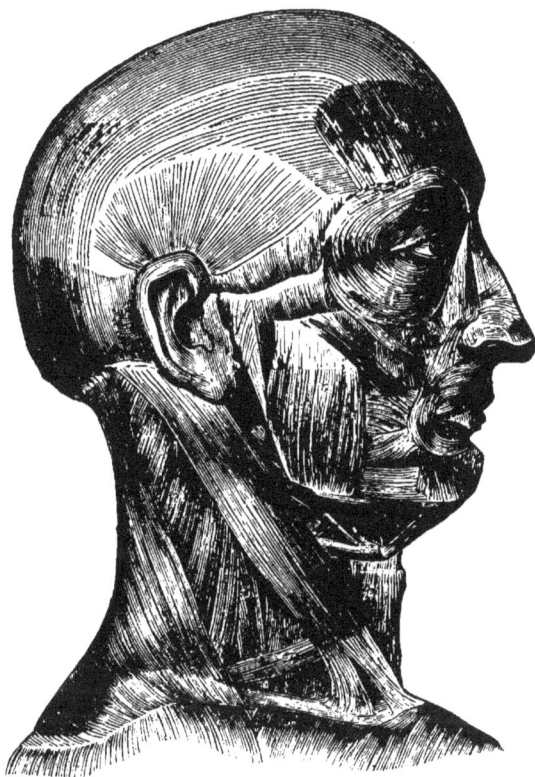

FIG. 15.—MUSCLES OF THE FACE AND NECK.

exclusion of that proper to the rest of the week. This over-exercise must be avoided under all circumstances. The modes of exercise must be left to each individual judgment and choice. Walking, in moderation, is perhaps the most convenient form, although riding calls additional muscles into play, and stimulates the nervous system by the peculiar open-air exercise attendant upon it. There should be but little if any fatigue after exercise, whatever its form may be. Some of the games in which children indulge to excess should therefore be practised with moderation, such as jumping rope, etc. It seems to be a delusion with many persons that the young can accomplish without injury feats of agility which those older in years would shrink from. Exercise in the open air is, however, strongly to be commended, but each one must learn what is his own capacity, and not exceed it. The object should be to promote the general health and develop all parts of the system in accordance with such a result, and not any part of it to excess at the expense of other parts or without regard to the improvement of the whole system.

CLASSIFICATION OF BONES.

I. THE HEAD, 29 bones.

SKULL—8 bones, viz:

1 frontal (*frontale*, frontlet).

2 parietal (*paries*, wall).

2 temporal (*tempus*, time).

1 sphenoid (*sphenos*, wedge).

1 ethmoid(*ethmos*,sieve)[head).

1 occipital (*occiput*, back of

*EARS—6 bones, viz:

2 malleus (mallet).

2 incus (anvil).

2 stapes (stirrup).

FACE—14 bones, viz:

2 lacrymal (*lacryma*, tear).

2 nasal (*nasus*, nose).

2 molar (*mala*, cheek).

2 turbinate (*turben*, whirl).

2 palatal (*palatum*, roof mouth).

1 vomer (plowshare).

2 superior maxillary (*superior maxilla*, upper jaw-bone).

2 inferior maxillary (*inferior maxilla*, lower jaw-bone).

BACK OF MOUTH—1 bone, viz:

1 hyoid (v-shaped).

II. TRUNK, 57 bones.

SPINE—24 bones, viz:

7 cervical vertebræ (*cervix*, neck). [back).

12 dorsal vertebræ (*dorsum*,

5 lumbar vertebræ (*lumbus*, loin).

SHOULDERS—4 bones, viz:

2 scapula (shoulder blade).

2 clavical (*clavis*, key).

BREAST—1 bone, viz:

1 sternum (*sternon*, breast).

*Two additional bones in a child's ear.

RIBS—24 bones, viz:
14 true ribs (attached to sternum).   [sternum).
10 false ribs (not attached to
PELVIS—4 bones, viz:
2 nameless bones.
1 sacrum.
1 coccyx (*cuckoo*).

III. UPPER EXTREMITIES,
60 bones.

ARMS—6 bones, viz:
2 humerus (upper arm).
2 ulna (elbow).
2 radius (spoke).
HANDS—54 bones, viz:
16 carpal (*carpus*, wrist).
10 metacarpal(palm).[thumbs).
28 phalanges (fingers and

IV. LOWER EXTREMITIES,
60 bones, viz:

LEGS—8 bones, viz:
2 femur (thigh bones).
2 patella (knee pan).
2 tibia (shin bone).
2 fibula (clasp).
FEET—52 bones, viz:
14 tarsal (*tarsus*, instep).
10 metatarsal (foot).
28 phalanges (toes).
SESAMOID—8 bones, viz:
4 in thumbs.
4 in great toes.

V. TEETH.
MILK TEETH—20:
8 incisors.
4 canines.
8 molars.
PERMANENT—32:
8 incisors.   8 bicuspids.
4 canines.   12 molars.

QUESTIONS.

What useful purposes do the bones serve?
What organs are protected by them from injury in the skull?
In the chest?   In the spinal column?
What organs are directly concerned in the movements of the body?
Of what two parts are bones composed? What is the effect of each?
What is the condition of bone when being first formed?
What useful purposes does cartilage serve?
How are the bones of the skull adjusted?
Is any one of the skull-bones moveable?
How are bones divided?
What is the interior structure of a long bone?
What is the effect of such an arrangement?
What is the appearance of the solid portion under the microscope?
What use have these canals?
What is the technical name of the back-bone?
What is its general arrangement?   How many bones compose it?
In what is the spinal cord contained?   [are its uses?
What is the arrangement of elastic tissue in the spine, and what
How can this be illustrated by experiment?
To what bones is the spinal column attached above and below?

What familiar nautical illustration is offered?
How does the head move on the spinal column?
What are the two upper bones of the spinal column called?
What is the thorax, and with what function is it concerned?
How many bones compose the upper limbs? The lower limbs?
How are the bones all attached to one another?   [animals?
What peculiarities in the limbs are found in the lower classes of
What is the membrane covering bones called? What are its uses?
How are bones connected together?
What kind of joints or articulations do we find?
What are ligaments? What is a sprain?
With what material are joints sometimes strengthened?
What fluid is found in the joints? What are its uses?
How do the various joints, as of the elbow, shoulder, hip, etc.,
differ from one another?
What organs preserve the erect position of the body?
What is the appearance of muscles under the microscope?
What kind of tissue separates the fibres?
What is the action of muscles wherever found?
Is this motion under the control of the will?
What is the arrangement of the fibres in voluntary muscles?
What great organ is an exception to this?
What is the arrangement of the biceps muscle, and how does it
connect with the bone?
What is a tendon? What is the most powerful tendon in the body?
What is an aponeurosis?
What is the result of contraction of a muscle?
What takes place when this contraction ceases?
What are antagonistic muscles? What are flexors? Extensors?
What is the effect of their separate action? Of their combined action?
What rest does the heart have from labor?
What rest do the muscles as a class have during their work?
What is the general muscular action involved in walking?
What is the essential point so far as the hip is concerned in this
What occurs during running?   [process?
What is the muscular movement in leaping? In swimming?
What effect has exercise, such as occupation, on muscles?
What effect have age and sex on their development?
What is the effect of gymnastics?
How is this exemplified in the training of horses?   [sleep?
How are the muscles affected in standing, sitting, etc.? During
What are flexor muscles? Extensor muscles? Abductors? Ad-
ductors? Depressors? How are the muscles arranged in pairs?
What peculiar action have the muscles of the face?
What duty is performed by microscopic musular fibres?
What muscles of the face assist in giving expression?
How many muscles control the motions of the face and neck?
What is the effect of the non-use of muscles?
What rules should be observed as to exercise?
What modes of exercise are commended?
What is the object of exercise?

DIGESTION.

Digestion is the process by which food becomes converted to the needs of the system. The changes which it undergoes occur mainly in the stomach and intestines. When taken into the mouth it is not in a fit condition to be absorbed, and therefore requires, at various portions of the digestive apparatus, such action upon it as will adapt it to the nourishment of the animal. The system is constantly undergoing a process of wear and decay, and digestion is the most important means of supplying materials for the restoration of life and vigor by the repair of such losses. When food is taken into the mouth, all the nutritious portions are separated from it as it passes through the different parts of the digestive apparatus. In other words, **Digestion** fits it to be absorbed, while **Absorption,** which we shall afterward describe, carries it into the current of the blood, through which fluid the body is directly nourished.

Digestion in the Vegetable.—Digestion takes place in the vegetable as well as the animal. It is of a very simple nature in the former. The soil supplies the materials, which are mainly water, carbonic acid, ammonia, etc., derived from the surrounding atmosphere. These are condensed in a porous soil and placed within the grasp—if we may so call it—of the plant, the rootlets of which become thoroughly mixed with the soil as they penetrate it in all

5   D

directions. The greater part of the carbon in the vegetable is derived from the soil in this way. Sometimes chemical substances, such as gypsum, which is a preparation of lime, are added to make the soil more porous, and thus increase its powers of condensation; charcoal acts in the same way. The various substances necessary for the nourishment of the vegetable come in contact with the expanded extremities of the rootlets; and under the influence of absorption, in the form of imbibition, as this is technically called, fresh portions are continually passing up to the stalk, forming the sap, which goes to the nutrition of the plant.

Food of Man.—As a rule, the articles employed for food are derived from the animal and vegetable kingdoms. By the term food, or *al'iment*, we denote such solid substances as are capable of being converted into *chyle* (pron. *keil*) —the digested mass after it has been subjected to the action of the alimentary canal—and such liquid substances as can be readily absorbed for the wants of the system. In some parts of the world, however, mineral substances, such as earth, are occasionally eaten alone or mixed with bread, but no nutriment is derived from such articles. They fill the stomach and satisfy its demands for the time being, but have no other effect. In countries where grain and fruit are abundant the inhabitants naturally adopt that form of diet, while in other regions, with little vegetation, they live almost entirely on animal food.

As already stated, the taking of food supplies materials to make up for the losses which the body necessarily sustains as a part of its existence. In early life the body gains in this way more than it loses, and hence it grows. After a while the amount lost and gained becomes about equal, and the body gains or loses but little in weight. Sometimes in old age the body loses more than it gains, and the weight decreases.

To ensure perfect health the same ingredients should be taken in the food that are necessary to the composition of the body. The substances usually employed as articles of diet, such as meat, milk, water, and vegetables, contain a variety of elements fully adapted to the wants of the system. Milk alone contains a sufficient number of these to form the sole diet of the young child, and indeed of the young of other animals.

Milk and flour are strikingly like the blood in their composition, and are therefore well adapted as articles of food for the repair of the body. This may be shown at a glance in the following table :

FLOUR.	MILK.	BLOOD.
Fibrin, Albumen, Casein, }.................Curd or Casein..	{ Fibrin. Albumen. Casein. Coloring-matter.
Oil.........................Butter...............Fat.
Sugar, starch.................Sugar...............Sugar.
Inorganic substances, such as preparations of soda, potassa, lime, iron, etc.	Inorganic substances, etc.	Inorganic substances, etc.

It will be seen that flour, milk, and blood all have sugar and fatty matter and chemical and albuminous substances in common.

Water, which is so necessary as part of our nourishment, contains a certain amount of mineral matter, and every portion of the body requires that mineral matter be present in its solids and fluids. Indeed, the animal begins to fall off in its nutrition when deprived of these materials.

Condiments.—Common salt is found in every part of the

body, and man and animals generally instinctively recognize
its necessity as an article of diet. When taken by man as
an aid to digestion it is called a *con'diment*, a name under
which are embraced pepper, mustard, and other such aux-
iliary articles. It has been thought by some that salt has
the power of assisting the activity with which certain arti-,
cles of diet are absorbed during the process of digestion.
Lime, which exists in the bones and teeth, and minutely
in other tissues and fluids, is taken into the system, often
without our knowledge, as an ingredient of water and other
articles; and the same may be said of other substances.
Soda, potassa, and magnesia are also found in different portions of the body. In studying the circulation of the blood
we shall find that iron exists in the coloring-matter of that
fluid. When this material is deficient in quantity the
individual acquires a peculiar paleness, and the general
nutrition of the body suffers. The natural color returns
under the use of iron as a medicine. The amount of iron
in the body has been estimated at about forty-five grains.

Ingredients of Food.—A number of articles employed as
food contain as their chief constituents the two elements,
carbon and hydrogen. These are starch, gums, mucilag-
inous materials, cane-sugar, and grape-sugar. Starch is
eaten in the form of potatoes, arrow-root, rice, corn, beans,
etc. Carrots, melons, turnips, cucumbers, etc. furnish cane-
sugar, which is also derived from the sugar-cane or beet-
root. Fruits, honey, etc., and wine, beer, cider, etc.—fermented liquids—contain grape-sugar. In some of the muscles, as of the heart, a similar sugar is found to exist. and
a kind of starch formed in the liver becomes converted also
into grape-sugar. Fats are usually taken as food in the
solid form, as butter or lard, or in the liquid, as oils, such
as olive oil. A classification of aliments has been made,

called the ni'trogenized, because they contain nitrogen. They are also called albuminous substances, because they contain albu'men, or matter resembling white of egg. These are usually derived from animals, being contained in the muscular parts of flesh, in the milk, in the white of eggs, and the gelatinous principles obtained in soups from the skin, bones, etc. of animals. Vegetables also supply them from the gluten or nutritious part of grain and seeds, of peas, beans, etc.

In addition to these essential articles, others are taken, according to individual caprice or fancy, such as condiments, already alluded to; alcohol in the form of wines, beer, spirits, etc.; tea, coffee, chocolate, etc. Some of these, as alcoholic drinks, injure the digestive process, even when taken in small quantity, and they may induce the habit of drinking. Acid drinks, such as lemonade, are sometimes used to relieve thirst, and pepper and mustard to stimulate the action of the stomach, but healthy digestion can go on perfectly well without them. When the habit of taking these articles becomes fixed, the tone of the stomach suffers from the excessive stimulation.

Classification of Food.—Some authorities have divided all food into two classes: those which contain nitrogen are called *nitrogenized,* and those which do not, the *non-nitrogenized;* the former predominating in animals, the latter in vegetables. A due admixture of both is necessary to perfect health. Nitrogen forms an essential constituent of the animal tissues; hence its importance as an article of diet. It has been shown that animals and vegetables contain both classes of food, so that animals which live entirely upon flesh could be nourished for a while on vegetable food, and *vice versâ.* As there is less nitrogen in vegetable food, an animal living entirely upon it is com-

5 *

pelled to eat a larger quantity in the course of a day than one which eats meat entirely. It is said that a dog will thrive if it takes one-thirtieth of its weight in animal food, while a horse consumes a tenth or twelfth of its weight in vegetable food. Animals that are restricted to one class alone soon fail in health and die, while they would live if confined to a single article, as milk, which contains both nitrogenized and non-nitrogenized elements. Variety of food is, however, essential to healthy digestion in all animals which are accustomed to a mixed diet. A disease called scurvy may result in man when such variety is impossible, as in some of the voyages of discovery to the Arctic regions; although here it was found, in more than one instance, that a change to another article of the same class relieved for a time this condition. All diet-tables for the army and navy, for the use of officers and men for the preservation of their health, are made up of a due admixture of both kinds of food.

Some of the more important ingredients of food, such as oils, fats, starch, sugar, etc., deserve brief mention.

Oils and Fats.—Animals and vegetables—such as the olive and the palm—supply oily matter of different kinds. This is generally found to consist of three chemical substances, called o'lein, ste'arin, and mar'garin. These three substances are compounds of glycerin, which is the sweet principle of fats, in combination with a fatty acid. An oily substance is soft or firm according to the proportion of these contained in it; it is called a *fat* if it is of firm consistence, and an *oil* if it is more fluid. When examined under a microscope, each drop of oily matter appears in round globules of different sizes, and the crystals of fat are seen mixed with them. Oils are not soluble in water, but they mingle with it in a mixture called an *emul'sion*, which is only a mode of holding oily matter in a state of fine division. Milk itself is an example of an emulsion, for under the microscope we find numerous oil-globules in it, and we obtain butter from it. Oily matters are largely taken as articles of food in regions of extreme cold, such as the Arctic regions, where it becomes necessary to keep up the temperature of the body by such diet. Tallow candles and the coarsest oils have sometimes been eaten in large quantities by the people of those latitudes.

Sugar.—This is derived from a variety of sources, such as the sugar-cane, sugar-maple, and beet root. It is present in the vegetable during inflorescence or flowering, and is the characteristic ingredient of sweet fruits, where it exists in combination with vegetable jelly. The bee collects it from various flowers and converts it into honey. Sugar is a very nutritious article of diet. It is soluble in water, ferments very easily, and is very digestible. Cane-sugar is the form usually employed, but in fruits another variety exists, called grape-sugar, which is not so sweet or so soluble.

Starch.—This element of food is very extensively diffused through the vegetable world. In rice, barley, etc. it is in an almost pure state, but it is mixed with sugar in some grains, as oats, and in seeds, such as beans, peas, chestnuts, potatoes; with mucilage in oily seeds, like the almond, hempseed, hazelnut, etc. Starch is prepared artificially from various substances, as wheat, corn, and the potato. From a hundredweight of potatoes we can get about fifteen pounds of starch. The different flours are largely starch. Starch does not dissolve in cold water, and is not, therefore, digestible in that shape, but under the action of boiling water the grains of starch discharge their contents and undergo solution.

Gum.—This is also found in vegetables, and is often used as an article of diet. During the gum-harvest season in some parts of Africa the natives are said to live entirely on it, taking about half a pound a day each. It is not considered, however, as specially nutritious, and is not very digestible. Gum arabic is one of the best-known articles of this class, and is itself the juice obtained from making incisions in a tree, particularly in the part of the world which has given it its name.

Albuminous Substances.—These are the most important of all the articles employed as food, without which life cannot be sustained. The nutritive principles of this class are called albumen, fibrin, and casein. *Albu'men* exists in its purest state in the white of egg, which gives it its name (*albus*, white). It is found in meat and milk and in most of the animal solids, and is present in considerable quantity in all the fluids which are concerned in the nutrition of the animal. The egg of the bird contains hardly any other ingredient except the albumen, if we except the yellow fatty matter of the yolk, and yet the whole chicken is developed from it, including feathers, claws, bloodvessels, and everything else that forms part of the animal. *Fi'brin* is found in the body in a fluid state in the blood and in the lymph and chyle—fluids which will be studied hereafter—and in the muscles. In many of the cereals, as they are called, or vegetable grains—in wheat especially—a highly nutritious substance is found, which is called *glu'ten*. It is obtained from wheaten flour by kneading under a stream of water. We see a good example of gluten in bread, which owes its peculiar porous character and consistence to that substance. Gluten contains vegetable fibrin and albumen, which resemble animal fibrin and albumen. *Ca'sein* is the principal ingredient in the milk, of the upper classes of animals especially, and is the envelope which surrounds the oil-globules or butter-globules floating in that fluid. What is called curds and whey is nothing more than the casein of milk coagulated with rennet, the rennet itself being an infusion of the mucous or lining membrane of the fourth stomach of the calf.

The albuminous substances are all rich in nitrogen, but to ensure perfect digestion it is necessary to combine with them a proper mixture of fatty principles or substances containing mineral matters, such as common salt, etc. Such a mixture is adopted almost as a matter of course; for example, bread and butter, potatoes and beef, or milk alone, which is itself a mixture of both albuminous and fatty matters with sugar, etc.

The following table exhibits the amount of starch, sugar, albumen, water, etc. in the various articles of food in 100 parts of each:

	Water.	Albumen.	Starch.	Sugar.	Fats.	Salts
Bread	37	8.1	47.4	3.6	1.6	2.3
Wheat flour	15	10.8	66.3	4.2	2.	1.7
Oatmeal	15	12.6	58.4	5.4	5.6	3.
Rice	13	6.3	79.1	0.4	0.7	0.5
Potatoes	75	2.1	18.8	3.2	0.2	0.7
Peas	15	23.	55.4	2.	2.1	2.5
Milk	86	4.1	......	5.2	3.9	0.8
Cream	66	2.7	......	2.8	26.7	1.8
Cheese	36.6	33.5	......	......	24.3	5.4
Beef	51	14.8	......	......	29.8	4.4
Pork	39	9.8	......	......	48.9	2.3
Poultry	74	21.	......	......	3.8	1.2
Whitefish	78	18.1	......	......	2.9	1.
Egg	74	14.	......	......	10.5	1.5

The study of such a table enables us to know what articles of diet should be taken together. Those that have plenty of albumen, for instance, should be combined with others that are largely starchy, in order to ensure a full diet.

Gelatin.—When the organic parts of bones, skin, tendons, etc. of animals are boiled for a length of time, they are converted into a substance called *gel'atin*, which is soluble in hot water, but cools into a jelly-like mass. Experiments have been made to determine whether it has any nutritious effect when given as an article of diet. It was said that when animals were restricted to this exclusively they fell off in their nutrition, and were compelled to have other kinds of food given to them as well. It has been found, however, that such might be the case if an animal were restricted to any one kind of food, whether it be gelatin or any other article.

We have thus seen that the important classes of articles used as food belong, as a rule, to the three great divisions of albuminous, oily or fatty, and sugars or starches, for the latter are convertible into sugar.

Animal and Vegetable Food.—If we consider the two great classes of animal food and vegetable food, we find much to interest us.

Animal Food.—Almost every part of animals has been used as aliment, but the portion universally employed is the *flesh*, by which we mean the muscles of voluntary motion. The use of blood as food was forbidden among the Hebrews—not from any injurious effect it might be supposed to possess, but because it was thought that they might by familiarity with it feel less horror at shedding the blood of their fellow-creatures. Attention has of late years been attracted to the presence of a minute animal—a worm, called the *trichi'na spiral'is* (which two words only' mean a "spiral hair-like animal")—in the various forms of pork, as ham, sausages, etc., especially when not thoroughly cooked. This worm may thus get a foothold in the human system and produce a painful and fatal disease.

The character of animal food differs according to its age, the material it has fed upon, on climate, season, fatness, etc. An extremely young animal's flesh has more gelatin in it and is more soluble, but not always on that account more digestible. Calves' meat—veal—is not as digestible, for example, as beef. The fat of the young animal is arranged around the muscles, while in the older it lies more between the fibres of which the flesh is composed, giving it a marbled appearance. So far as the season is concerned, the flesh of animals is in better condition during the early winter months after an abundant feeding during summer; in the spring the flesh is lean from deficiency of nourishment. Fat animals are more desirable for food, because the admixture of fat and lean makes the flesh more soluble. The lean of fat meat is therefore to be preferred on account of its being nutritious and digestible, although the fat itself is less digestible than the flesh.

Meats and game that have been killed a few days before being cooked are by many persons considered more tender; but they should not be kept for so long a time that they begin to be decomposed or tainted. The use of half-spoiled meat or game is certainly to be condemned as supplying to the system materials which are not fitted for digestion, or which may disturb the action of the stomach and bowels. The odor connected with the decomposing of such meat is usually a sufficient warning.

All parts of birds, as well as the eggs of many of them, have been taken as human food. Those parts which are continuously exercised are not as tender; hence a chicken's wings are more delicate than the legs. A duck's legs, being less often used, are quite as tender as its wings. A woodcock's wing is more tough than that of the partridge, because it flies more and walks less than the latter. Birds that chiefly live in the water, as the goose and duck, are not, as a rule, so digestible; their fat is often rancid and fishy.

The flesh of *fish* is much used as an article of diet. It consists mainly of fibrin, albumen, and gelatin. It is less nutritious than the meats usually employed, but more so than a purely vegetable diet. An oily fish, such as the eel, is less nutritious, as the oily matter is not so easily digested in the stomach. Tribes are found in various parts of the world who live on fish alone, that being the food which they can easily secure, after floods or heavy rains, with least trouble and expense. Such are said to be *ichthyoph'agous*—a sonorous word, meaning, from its Greek derivation, "eaters of fish." Nearly every part of the fish is considered nutritious. A fish diet has been recommended by some as appropriate in cases of mental exhaustion and other forms of brain disturbance, because of the phosphorus contained in such food, the brain being largely supplied with that stimulating chemical substance, and being supposed to receive a fresh supply when this kind of food is taken.

Milk.—As already stated, milk is well adapted from its composition to sustain life, being composed of such albuminous, fatty, and saccharine matters in water as are well calculated for the nourishment of the body. When taken into the stomach it is doubtless converted by the fluids of that organ into a solid curd and a fluid or whey. The curd is digested just as any other substance would be digested there, while the fluid portion is absorbed into the blood-vessels, and soon becomes a part of the blood itself, going to the nourishment of all the living organs. Its nutritive properties adapt it to both sick and well. The cheesy **part**

of the milk sometimes disagrees from its not being digestible, and the curd formed in the stomach may remain there for hours. Generally milk agrees better than any other kind of food. Boiling makes milk much more digestible, especially if a little lime-water be added to it should it produce an acid state of the stomach.

Derived from milk are cream, butter, cheese, buttermilk, and whey. *Cream*, carefully skimmed from the milk, contains casein and whey, together with the butter. Its oily nature renders it less digestible, but it does not turn acid as easily as milk. *Butter*, which is the oily part of milk, obtained by churning, is nutritious, but, like other oils, not very digestible. *Cheese*, which is the curd of milk, pressed and partially dried, carrying with it a little butter enveloped in it, is very nutritious, although not very digestible. It is considered by some in the light of a condiment, for, according to an old proverb,

> "Cheese is a surly elf,
> Digesting all things but itself."

Cottage-cheese, or smeer-case—the soft curd of milk—is more easily digested. *Buttermilk* is the fluid left in the churn after the removal of the butter. When prepared from milk that has only been kept a little while, it may be perfectly sweet when first obtained. Having really lost the greater part of its fatty matter by the churning process, it should be more digestible, though not so nutritious. *Whey* contains a little sugar, cheese, and butter, which render it slightly nutritive, but it is apt to turn acid in the stomach.

In Tartary, especially, the people use largely a beverage which they call *koumyss*, a kind of sour buttermilk, prepared from milk that has been allowed to stand for some days in a leathern churn until it becomes sour, when it is bottled for use.

6

Eggs.—These have already been alluded to in referring to albumen. They are more nutritious than milk, but less digestible, no matter how they may be prepared. The white of the egg is almost wholly albumen in a pure state; the yolk is oily matter with some albumen. Like albumen generally, the white of the egg coagulates by heat. It is more digestible when slightly boiled than when raw. The yolk, on account of its fatty matter, is not so digestible as the white. One great advantage of the egg as an article of diet is the large amount of nutritive matter contained in a small space. Fried eggs are less digestible, for the albumen becomes hardened and the oily matter is altered in its nature by the heat, so as to be unfitted for digestion. It may be stated, while referring to eggs as articles of diet, that all substances that contain a large amount of nutritious matter in a small compass are not easily digested by the stomach.

Vegetable Food.—In some portions of the world vegetable food is almost the exclusive diet, and there are many animals—those we shall hereafter describe as the herbiv'orous class—which live entirely on this kind of food. As the great difference between animals and vegetables, in their chemical constitution, is the presence of nitrogen in the former, the question naturally arises, How does the flesh of the animal get the nitrogen of which it is composed if there is none in the food? The air it breathes is largely made up of nitrogen, and this is doubtless the chief source of its supply, passing into the animal's lungs, and thence into its general system.

Bread is usually made from the flour of wheat. It contains principally a saccharine or starchy matter and gluten, to which reference has already been made. When prepared with yeast as a ferment, or with old paste or leaven kept

for the purpose, it is said to be *leavened,* and *unleavened*
when made without such an addition to it. The effect of
a *ferment* is to produce a thorough change of character and
decomposition of the whole mass in which it is placed,
and such a result takes place in regard to the flour. We
shall hereafter see that the presence of a ferment, called
pep'sin, is necessary in stomachal digestion. Milk sours
and becomes curdy if kept for a while in a warm place,
for the casein in it becomes itself so much altered that it
acts as a ferment for the whole mass, developing lactic
acid, which coagulates it. So with the flour; the gluten in
it, mixed thoroughly with the starch and water, produces
what we call bread. Unfermented bread will, of course,
be more acceptable to the stomach, as the bread prepared
by the other process is often somewhat acid. Sweet cakes
are apt to be much less digestible, for they are generally
made up of sugar and eggs; and the same objections hold
good in regard to pastry.

Breads are made also of other flours, as bran, rye, barley,
etc., but they are not so digestible or nutritious as that prepared from wheat, and are apt to turn acid on the stomach.
Rice and corn flour are also eaten as bread, and are both
digestible and nutritious. When soup is taken as an article
of diet, the addition to it of any kind of bread, preferably
wheaten bread, gives the soup a digestive texture, and enables
the gelatin in it to be much more readily digested. The
difference of texture of bread when hot and fresh and
spongy, and when stale, probably accounts for the latter
being more digestible.

Fruits.—By some these are considered as the most healthful articles that can be taken into the stomach; by others
they are looked upon as a fertile source of stomach derange-
ments. As a rule to be generally observed, fruits should

be eaten only when perfectly ripe, the seeds and skins, which are thoroughly indigestible, being carefully removed. Such fruits as the melon and canteloupe are the least digestible of all this class. Preserved fruits act on the digestive organs just as sugar itself does, although the combination of fruit with sugar is not so likely to disorder the stomach as either sugar or fruits alone.

Cooking.—The value of food depends upon its digestibility and upon the amount of materials it contains adapted to the nourishment of the body. The manner of preparation of the food in cooking is also an important consideration. Few articles of diet are ready at once to be taken into the stomach without some such process. Cooking imparts flavor, and thus excites the secretions of the mouth and stomach, and at the same time renders the article more digestible by the softening and division of its particles, mechanically or by the aid of heat, water, or condiments. By heat especially, as in the making of beef-tea, the nutritious parts are entirely separated.

The methods of cooking, such as roasting, boiling, stewing, etc., differently affect the digestibility of food by modifying its flavor, the arrangement of its particles, etc. It is said that the internal temperature of cooked meat should never go beyond 160°, nor fall below 130°.

In *roasting* the heat is applied directly to the substance, which is placed in front of the fire or in an oven. Animal food is affected by roasting as follows: the albumen is coagulated, the fibrin shrivelled up, the fatty matters melted, and some of the water driven off. The appearance of meat after roasting is too familiar to be described. Of course, after losing so much fat and water, the weight is greatly lessened, but what is left in the meat is full of nourishment, while the gravy is also rich in nutritive material.

Broiling is a means of applying heat directly to the article of food by placing it before or over the fire, as on a gridiron. The effect is to brown the surface and to retain the juices of the meat, and this process is regarded as the best to preserve its tenderness.

Boiling has the effect of softening the food, so that the stomach can more readily act upon it. Water or steam is the vehicle through which the heat is applied. The boiling water dissolves out some portions of the flesh that are retained by the other processes, and about eighty per cent. of the salty matters, and the heat coagulates the albumen, while the gelatin in the meat is converted into a jelly-like mass. If there be much albumen in it, boiling to excess will produce a hard, over-boiled mass; while if there be much gelatin, as in young meat, a jelly-like substance results, which is quite as indigestible. Boiling produces a loss of weight, but not nearly so much as by roasting. In *stewing* a less amount of water is employed.

Frying not only requires a pan in which the article is placed, but also the addition of some fatty matter. As oil or fat is thus exposed to heat, and mixed with the substance, it is less digestible, and is very apt to disagree.

When vegetables are cooked, some parts, as the gum and starch, are dissolved out; if not soluble, they are softened, and thus better fitted for digestion. Some vegetables require much boiling to make them either palatable or digestible. Fruits which might disagree if eaten raw may be roasted or stewed or baked, and their fibres placed in such a condition by these processes as to render them digestible.

Drinks.—The drinks used by man are either water alone or in solutions of various kinds, such as alcoholic wines, liquors, etc.; gaseous or aërated, such as mineral waters;

6 *   E

acid or saccharine; and infusions, as they are called, such as tea and coffee, which are mainly boiling or cold water in which these substances have been placed for a brief period of time.

There is no fixed rule to be laid down as to the amount of drink necessary for an individual to take in the twenty-four hours. Much depends on habit, and some persons can drink quarts, while others, equally healthy, use but little fluid. Whatever the system requires must be taken. Sometimes solid food is swallowed so rapidly that the saliva of the mouth, to which we shall presently refer more minutely, cannot soften it. It therefore becomes necessary to swallow a considerable amount of fluid, such as water, to produce this effect. Should such be the case, or in any event, the amount of water or other fluids swallowed at meals should be moderate, not excessive, as the food becomes too much softened, and the secretions of the stomach so much diluted as to be impaired in their efficiency for digestive purposes. Acid and sweet drinks, or drinks of any kind, should be avoided immediately before meals, as they unduly excite the secretions of the stomach before food, the natural stimulus, is taken. Hot drinks stimulate the stomach to increased secretion, and at the same time increase the action of the muscular coat of that organ; but habitual stimulation of this kind is injurious, and must finally weaken it.

Water.—To be suitable for diet, water should be fresh, without smell, and comparatively tasteless. It should contain a little gas or mineral matter, but without any organic matter, such as animal or vegetable life. When taken into the body, as in drinking, it replaces what is lost by evaporation from the skin, from the lungs, etc.

Of the waters used for drinking purposes, *rain water* should be the purest if properly collected; that is, not from the roofs of houses, where impurities are apt to exist. It is much more insipid than river water, even when pure. *Spring water* may contain, in addition, a preparation of lime. If it forms a curd with soap, it is called hard water; if not, it is called soft water. The latter is preferable for domestic uses, as it has more effect in dissolving vegetable substances, and is not likely to produce any chemical changes. It contains more air than river water, and is therefore more sparkling. *River water*, being obtainable in large quantities, is generally used for drinking purposes, but is liable to be rendered impure by matters emptied into it. *Well water* may also be injured by impurities filtered through its walls, but a properly lined well will protect itself from them. *Lake water* is likely to be unfitted for drinking purposes, for into it are poured impurities from springs and rivers, such as decayed animal and vegetable matter, which may have remained in stagnant waters until decomposed and then been emptied into the lake. Foulness in water may be corrected by filtering through gravel, charcoal, etc., or boiling before being filtered, and then shaking it to bring back to it some of the air it has lost. Impurities may also be driven out by distilling the water; that is, by heating it until it becomes vapor, and then as the vapor cools collecting it drop by drop in another vessel.

Wine is the juice of the grape after it has undergone fermentation. To produce a wine it is necessary that there be present sugar (or some substance that will form sugar, such as starch), water, and a ferment, all exposed to a temperature of 70° or 80°. Red wines are made from the juice of the black grape; white wines, usually from the white grape or the dark grape deprived of its outer covering. To produce effervescence the wines must be bottled before the fermentation is completed. The gas formed is

carbonic acid, and, being unable to escape, it remains dissolved in the wine until the cork is removed. Sweet wines
are made so either by the grape-juice originally having an
excess of sugar in it or from the addition of sugar to it.
All wines contain in varying quantities alcohol, water,
mucilage, sugar, coloring-matters, and mineral substances.
Some of the latter give them their tartness, while the alcohol gives them strength. The proportion of alcohol in
sherry and port wines is about sixteen to twenty-five per
cent.—much less, as we shall presently see, than brandy,
whiskey, and gin, all of which contain from fifty to sixty
per cent. Strong wines deposit in time a crust, which is
largely tartar, the loss of which makes the wine more
grateful to the stomach, so that old wine is really made
better by age. The alcohol and carbonic acid in the wine
are produced by the gradual action exerted on the sugar
of the juice of the grape by the albuminous part of the
grape, which has undergone fermentation.

The following table gives the percentage of alcohol in
the more common wines and spirits:

Port wine	16 to 23 per cent.
Sherry	16 to 25 "
Madeira	16 to 22 "
Champagne	5 to 13 "
Claret	9 to 13 "
Hock	6 to 16 "
Brandy	50 to 60 "
Whiskey	50 to 60 "
Rum	60 to 75 "
Gin	50 to 60 "

Malt Liquors.—There is much less alcohol in these than
in wines; but there is enough to intoxicate. A bitter narcotic principle is present, imparted to them from the hop.
It is better to avoid their use entirely, because they may lead

to the habit of taking alcoholic drinks. Beginning with cider or beer, it will not be long before the drinker will get into the habit of taking stronger liquors, as brandy or whiskey. Malt itself is barley which has been made to germinate by warmth and moisture, and afterward dried. Part of the albuminous matter of the barley is converted into a ferment called di'astase, which changes a great deal of the starch of the seed into grape-sugar or a gum called dextrin. Ale and beer are obtained by fermenting an infusion of malt.

Spirits.—When wines are distilled we procure spirits, such as *brandy*, which is obtained from the wine of the grape. *Whiskey* is obtained from grain—rye, maize, and wheat being used for the purpose. In the fermentation the starch is converted into sugar, and finally into alcohol and carbonic acid gas. *Rum* is obtained from fermented molasses. *Gin* is distilled from corn and juniper-berries, or from some substitute for them.

It is a strange fact that so many of the different drinks used in all parts of the world, such as tea, coffee, cocoa, etc., should contain almost identically the same active principle, with the same chemical composition. They differ in other qualities, for coffee has more aromatic matters in it, and tea has tannic acid, which makes it more astringent. Tea and coffee stimulate the nervous system, but are not followed by the after-depression characteristic of alcoholic stimulants. They therefore give prompt relief in cases of fatigue, especially the coca-leaves of South America, which the inhabitants of that region use for this purpose after excessive labor and wearisome marches.

Quantity of Food.—As regards the quantity of food which should be taken by any one in health, no exact amount can be mentioned, so much depends on the capacity and health of the individual, the age and sex, as well as the influence of climate. Excess is a great cause of ill health, especially if the food be hastily swallowed, and hence improperly prepared for digestion. It has been estimated by physiologists that the average quantity daily consumed by a person in full health is about a pound of meat, rather more than that quantity of bread, about a quarter of a pound of fat, and between three and four pints of water, either taken as such or in the form of milk, tea, or other fluids. A larger amount of food is consumed in cold than in warm climates, and in cold seasons of the year than in hot. It is said that the daily amount of food taken by an Esquimaux is twelve to fifteen pounds of meat, including a large mass of fat. The case is cited by a Russian admiral of an inhabitant of those far northern regions who ate, in his presence, twenty-eight pounds of boiled rice and butter at a single meal. Mention is made by one of the Arctic explorers of an Esquimaux who in twenty-four hours devoured thirty-five pounds of meat, in addition to a number of tallow candles.

Children and youth really require, in proportion to their size, more nourishment than those who are older; but the aged have often to take more in proportion on account of the failure of their digestive organs and their inability to get the full benefit of what they eat. Those who take exercise require more than those of sedentary habits.

Meals should be regular and far enough apart to allow of full digestion being completed. Some persons require much more time than others, but scarcely any more than four or five hours of interval. Children can eat at briefer intervals than the adult, but the quantity at each meal should be moderate, not excessive, so that the stomach will not be taxed beyond its powers. Supper should be light, and, as a rule, without animal food, especially in the case of children, as the circulation may be interfered with by a full supper, particularly before going to bed, and the brain be disturbed so much as to cause dreams or restlessness.

Change, rather in the nature than the quantity of the diet, is desirable. The strong require a different kind of diet from the weak, and the occupation or pursuit in life has also to be considered. Those who are engaged in heavy and fatiguing labor should eat more food and articles of a more substantial character than those who are devoted to literary work. Delicate persons and those with weak digestive powers should avoid articles that are not easily digested, and depend on light but nutritive diet taken in small quantities and frequently repeated. Bread and milk, with meat in moderation, and a small amount of vegetable food with fruits, form a good diet for such persons. As will be hereafter stated, food should not be hastily swallowed, as the action of the teeth in dividing the food, and of the saliva in softening it, will not then take place.

Conditions Necessary for a Healthy Diet.—These may be briefly stated :

1. The food taken must contain a proper proportion of the different principles that already exist in the human body.

2. It must correspond in quantity and quality with the amount of work performed by the individual himself to supply the matter and force spent by him.

3. It must have flavor or taste to render it palatable, and therefore more digestible.

In addition to these properties, it must not be forgotten that the influence of climate, such as temperature, affects the activity of digestion, as well as the quantity of food to

be taken.   In a cold climate exercise renders the appetite
brisk, and a larger amount of food is required; there is'
also an increased demand for internal heat, so that fatty
matters are necessary as food, which produce when oxidized
a large quantity of heat.

Appetite and Hunger.—Hunger is an internal sensation,
always referred to the stomach.   In its slightest manifesta-
tion it is simply *appetite*, cr a desire for food or drink.
It may be temporarily relieved by substances that are not
nutritious, or by emotion, and is increased by moderate
mental and physical exercise.   The sensation of hunger
occurs after the stomach has been for a while empty, digestion being accomplished.   Hunger arises from a real
want of the system.   It is more powerful in childhood and
youth, and also in old age after the digestive powers have
begun to give way.   The sensation seems to be the result
of a nervous impression made upon the stomach from the
needs of the other organs for nourishment.   When death
occurs from protracted hunger, the young and robust die
before those who are older, the activity of all the functions
being greater in them, and the necessity for food being
therefore more imperative.*   Hunger is an irresistible desire that cannot be disappointed.   If no food is taken,the

---

* Dante in the *Inferno* describes the sufferings and death from hunger
of the Count Ugolino, and shows a correct knowledge of the more serious
effects of abstinence from food on the young than on the old; although
on this latter point physiologists differ:—

  " Now when our fourth sad morning was renewed
   Gaddo fell at my feet, outstretched and cold,
  Crying, ' Wilt thou not, father, give me food ? '
   There did he die, and as thine eyes behold
  Me now, so saw I three fall, one by one,
   On the fifth day and sixth ; whence in that hold
  I, now grown blind, over each lifeless son
   Stretched forth mine arms."

tissues of the body break down, and the individual is compelled to live on himself, as new materials for the nourishment of the body, not being obtained from without, must
be derived through the blood from all organized parts of
the body. The result is, that the harmony of action of the
various organs is broken up; the functions of the body,
such as digestion, absorption, preservation of its heat, and,
finally, respiration and circulation, gradually cease, and
death takes place. The blood, changed in its character,
is not fit to nourish, the brain therefore suffers, and the
person thus affected often becomes delirious before death.
He may die of cold, for the food necessary to keep up the
temperature of the body will not have been supplied to it.

It has been supposed by some that hunger was caused by
the coats of the stomach rubbing against one another; but
this is not probable, as there can be but little friction between walls made up of such flexible membrane. Others
have imagined that hunger is excited by the presence of
gastric juice, as the secretion of the stomach is called, which
irritates the lining of that organ and gives rise to the desire; but this is not so, for experiments have shown that
during fasting no gastric juice is poured out in the interior of the stomach. Hunger is a vital action beyond
our powers to fathom, being the expression of an internal
want.

The vegetable exhibits something much akin to hunger,
and, if we may so call it, we must believe that this sensation can sometimes exist without the necessity of a nervous
system, for the vegetable is devoid of all nervous apparatus.
The plant being fixed in the soil, its rootlets go out in all
directions in search of the food which it requires. This
action seems to resemble the promptings of hunger in the
animal, for it is the evidence of a desire for food dependent

7

on the wants of the system. There must be a supply to take the place of what has been expended.

Thirst.—Thirst is a local sensation, referred to the throat, being a desire for liquid nourishment. It is characterized by a feeling of dryness of the lining membrane of the throat, caused by a positive want of fluid on the part of the system from a diminished amount of water in the blood from any cause, such as increased secretion of perspiration, etc.; from arrest of secretion of fluid from the lining membrane of the stomach or intestines; or from emotion. A certain amount of fluid taken daily is essential to perfect health. Thirst is more imperative than hunger, and death results from it more rapidly, as has been observed in the cases of the shipwrecked. Salty articles of food produce thirst, because they require a large supply of watery fluid for their solution, derived through the walls of the alimentary canal, and thus the amount of water in the blood is diminished. Thirst may sometimes be relieved temporarily by bathing, as in salt water, a sufficient amount of fluid passing through the pores of the skin. It is said that after the second day of deprivation of water the most terrible suffering ensues, and that death will usually result in from seven to ten days. It is not easy to state how long a human being can exist without drink, so much depends on the state of the system at the time, the condition of the atmosphere, etc. It is said that an animal will live longer on water alone than on any dry principles of food.

Apparatus for Digestion.—The simplest form of digestive apparatus is that of the vegetable, which receives its nourishment directly from the earth through its roots or by its leaves from the surrounding air. The food it thus obtains requires no preparation, as in the animal, which necessarily possesses an internal arrangement of organs adapted to the purpose. In the very lowest form of animal life a simple sac or cell, without any opening whatever, is all that exists for the entire digestive process; there is no stomach, for none is required; everything is at once absorbed.

In another form of low life the whole animal is a sort of sac, with only one opening, thus resembling a delicate gum-bottle. The animal is all stomach, and like a membranous bottle can be turned inside out; the former inner surface will soon become like the external, and absorption will take place from the new interior surface. In the infu'sory animal'culæ, as they are called, which are microscopic animals found in various fluids, the same general type exists, on a simple scale, as in man; there is a body having an opening for the reception of food, a dilated part or reservoir like a stomach, with an open canal leading from it. As we rise higher in the scale the apparatus becomes much more complicated, according to the nature of the food on which the animal lives. When the organs of digestion are developed to their fullest extent, as in man and the higher classes of animals, they include a cavity or series of cavities in which the nutritious portion of the food is prepared for absorption, and the non-nutritious portions of food are expelled. In man, the apparatus of digestion consists of a long canal, varying greatly at different parts in size and structure. It is usually divided into the following parts : mouth, pharynx, œsophagus or gullet, stomach, and intestines. The stomach and intestines occupy the greater part of the abdo'men, and are the chief organs interested in the process of digestion.

The *mouth* receives the food, which is here subjected to the action of the teeth and of certain juices which are poured into that cavity to soften and dissolve it. The tongue, cheeks, and jaw by their movements also aid in

producing changes in the size and character of the particles
of the food. The fluid poured into the mouth is called
the *sali'va*, which is formed by little bodies called *sal'ivary
glands*, the fluid being carried from them by small canals
or ducts passing from the glands and opening into the
mouth. This process is called *insaliva'tion*. The action of
the teeth in cutting, crushing, or grinding the food is called
mastica'tion. The food thus becomes finely divided or
comminuted. Connected with the jaws are muscles, which
are very large in some animals, as the horse, through which
additional force is exerted in mastication.

The Teeth.—The teeth are composed of a material like
bone, called the *dent'ine* or *ivory*, and are inserted deeply
into the jaws by roots or fangs. The portion of the tooth
above the gum is called the crown, which is covered by a
thin layer of enamel, the hardest material in the body. In
the interior of the tooth is a cavity containing nerves and
bloodvessels. The young child when it gets its first set of
small teeth complete—the *temporary* or *milk teeth*—has
twenty in all—four *inci'sors*, two *cani'nes*, and four *mo'-
lars* in each jaw. This early set is developed before birth,
but cuts through the gums at the following ages: incisors,
seventh to ninth month; canines, eighteenth month; molars,
twelfth to twenty-fourth month. When the child grows to
be six or seven years old their roots are absorbed, and the
teeth are loosened and fall out, to give place to the *permanent teeth*. These are thirty-two in number—sixteen in
each jaw—arranged in each half jaw as follows: two incisors, one canine or dog-tooth, two bicuspids (having two
points or cusps), and three molars or grinders (from *mola*,
"a mill"). The permanent teeth are developed in the jaw
below the milk teeth, and both are found at the same time,
about the fifth year of age. They gradually push out the

milk teeth from that time of life until the twelfth or four-
teenth year of age. The appearance of each of these sets
will be seen in the accompanying illustrations (Figs. 16, 17).

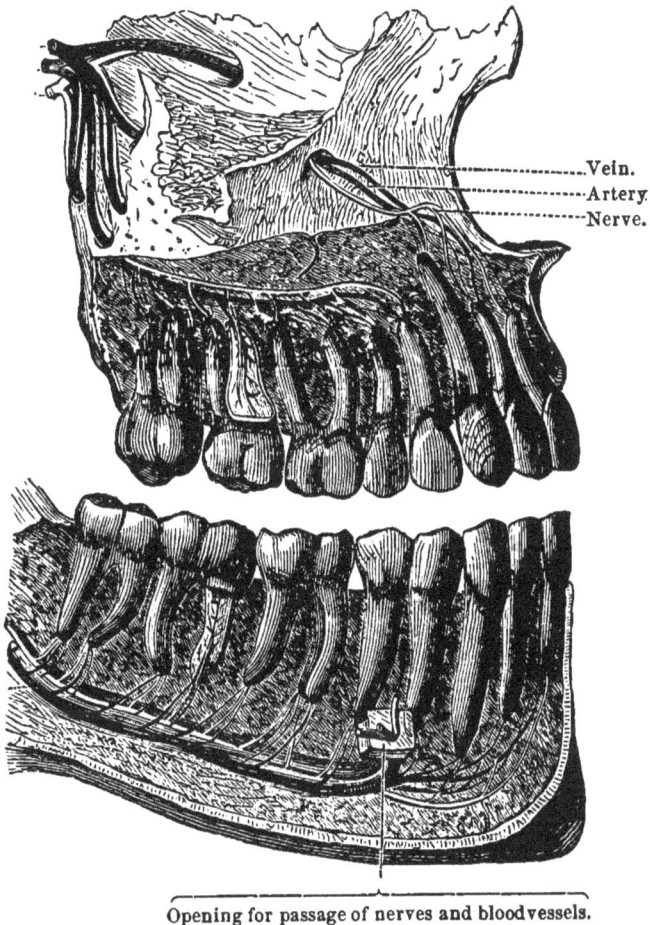

Opening for passage of nerves and bloodvessels.

FIG. 16.—GENERAL VIEW OF THE TEETH.

The "wisdom-tooth," so called on account of its not making
its appearance until the twentieth or twenty-first year of
life, is the third molar tooth of each jaw.

The teeth are of such shape in various animals as to be

7 *

adapted to the special food on which they live. The sharper teeth in front are called inci'sors, from their cutting properties, and the canine because they are like those of the dog. The motions of the jaw also admit of a cutting or crushing action upon the food. In animals which live entirely on animal food the jaws are very strong, and move more

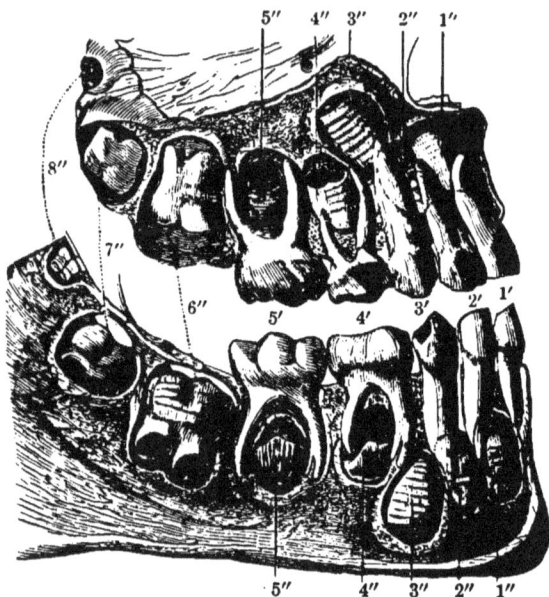

FIG. 17.—TEETH OF FIRST AND SECOND DENTITION.

1' to 5' are the teeth of first dentition, or milk-teeth; 1" to 8" are teeth of second dentition, or permanent teeth.

readily up and down than from side to side, as mastication is not necessary. The muscles of the jaw are also very strong, and the cutting teeth are much more decidedly developed than those for crushing, the latter being sharper than in other animals. By an inspection of a single tooth we can usually decide the nature of the animal, and the naturalist can even build upon this tooth, as a basis, a plan of the animal's general structure, habits, etc.

An animal which feeds entirely on flesh is called *carniv'orous* (*caro, carnis,* "flesh," and *voro,* " I eat "). One that eats grain is called *graminiv'orous;* one that lives on grass, *herbiv'orous.* Man, with teeth and digestive apparatus adapted to all kinds of food, is called *omniv'orous* (*omnis,* " all," and

FIG. 18.—TEETH OF INSECT-EATING ANIMAL.

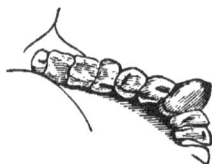

voro, " I eat "). Animals which feed entirely on insects (*insectiv'orous*) have conical-pointed teeth, which fit closely into one another in the two jaws like the teeth of clock-work (Fig. 18). Those animals which exist mainly on fruits (*frugiv'orous*) have the teeth rounded rather than sharp (Fig. 19). The grain-eating or *graminivorous* animals,

FIG. 19.—TEETH OF FRUIT-EATING ANIMAL.

whose food requires crushing or bruising rather than cutting, have large flat teeth, which, like millstones, crush and break up the food into fine particles by the side-to-side movement of the jaws (Fig. 20). These teeth have been called molars

FIG. 20.—TEETH OF GRAIN-EATING ANIMAL.

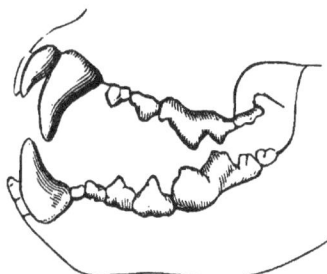

(from *mola,* " a mill "), on account of this very action. In the carnivorous animal (Fig. 21) the molars are sharper, and so placed as to meet like scissors-blades, while the canine teeth are largely developed. In the horse, the sharp front incisor teeth of other animals become more like the molars in structure and shape. Almost all animals have

FIG. 21.—TEETH OF FLESH-EATING ANIMAL.

molar teeth in some form or other. The teeth of the elephant are entirely molar. Some animals, as the whale, are
entirely devoid of teeth. In man, the teeth occupy an intermediate place between the carnivorous and herbivorous
animal; of his thirty-two teeth, twelve correspond to those
of the carnivorous and twenty to those of the herbivorous.
The movement of his jaws is upward and downward and
sideways.

The Tongue.—The tongue assists in mastication by en-
abling the particles of food to be worked around in the
mouth, to be acted upon by the various juices poured into
it. In some of the lower animals the tongue acts as a
means of seizing upon their prey, which is to them their
main source of nourishment. While man uses his hands
to convey food to his mouth, the elephant is provided with
an extension of the snout, called the trunk, which has
powers of suction with which it can carry articles into its
mouth. Insects have feelers around the mouth for similar
purposes. In some animals the motion of the tongue in
the mouth enables them to suck in liquid food, just as the
same process is effected in man. Juices are thus imbibed
by insects through tubes or suckers.

The Salivary Glands.—These glands (Fig. 22) are found
on each side of the mouth in man, and are three in number. They pour out a thin and a thicker kind of fluid,
which becomes mixed with that secreted by the membrane
lining the mouth. Although saliva is being constantly
formed and swallowed, the presence of food in the mouth
excites these glands to increased action, and an additional
quantity of saliva flows into that cavity. Even the mere
sight of agreeable food often has this effect, which is well
known as "mouth-watering." The amount of saliva secreted varies with the kind of food. If it is hard or dry

a larger quantity is poured out. It was found by experiment on horses that 400 parts of saliva were mingled with every 100 parts of hay, but only 50 parts of saliva were furnished when 100 parts of green stalks and leaves were taken. The saliva softens and moistens the food, and con-

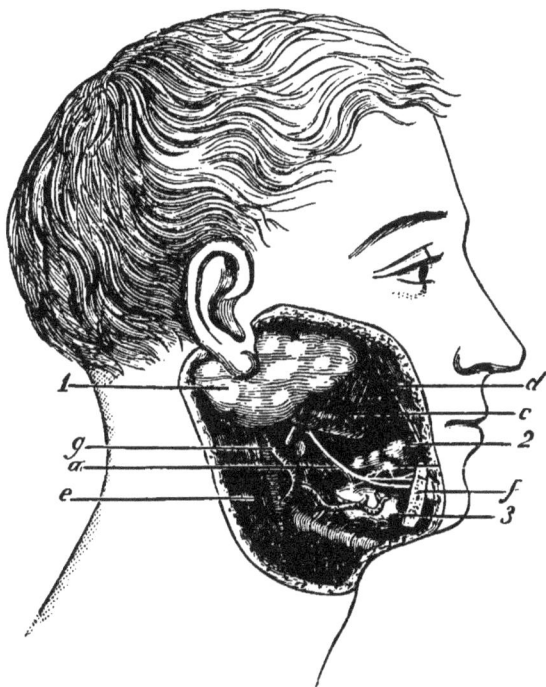

FIG. 22.—SALIVARY GLANDS.

1, parotid gland; 2, sublingual gland; 3, submaxillary gland; *a*, nerve; *c, d, e*, muscles of face and neck; *f*, lower jaw; *g*, artery.

verts the starchy matter contained in it into a gummy substance called dex′trin, and afterward into grape-sugar. This, being soluble, is more easily absorbed. It seems that this power depends on the admixture of the mucus of the mouth with the saliva. Perfect mastication, or division of the food, and perfect insalivation are necessary to ensure perfect digestion in the stomach. The structure of the salivary

F

glands, when examined under the microscope, resembles a
bunch of grapes. Their ap-
pearance is shown in Fig. 23.
The average amount of saliva
secreted in the course of twenty-
four hours has been estimated at
from two to three pounds. The
terms *parot'id*, *submax'illary*,
and *subling'ual*, applied to the
salivary glands, merely mean,
from their derivation, that they
are near the ear, under the jaw,
or under the tongue, as will be seen in Fig. 22.

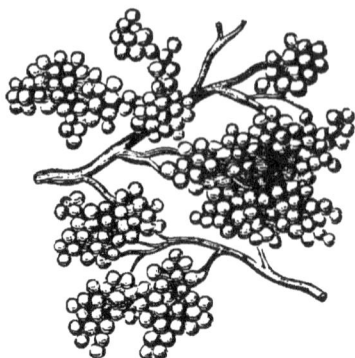

FIG. 23.—STRUCTURE OF A SALIVARY
GLAND.

Deglutition.—When the food has undergone these processes in the mouth it is swallowed. Swallowing, or *deglu-
tit''ion*, as it is technically called, includes the passage of the
alimentary mass from the back part of the mouth into the
stomach, and requires the action of the mouth, throat or
pharynx, and the œsophagus.

The pharynx and œsophagus together form a muscular
canal extending from the mouth to the stomach (Fig. 24).
The cavities of the mouth, nose, and larynx open into the
pharynx. The pharynx is separated from the mouth by a
fold of muscle and membrane called the ve'lum, or soft
palate. The œsophagus is about nine inches long, and,
like the pharynx, is lined with mucous membrane and supplied with a muscular coat. The first part of the act of
swallowing is under the control of the will, but after this
is accomplished the passage of the food from the throat to
the stomach is entirely involuntary. This is a wise provision, for otherwise the particles swallowed might pass
into the upper opening of the air-passages, as they occasionally do when, in common language, they " go the

wrong way." During the act of swallowing the larynx is elevated by the muscles connected with it, and removed out of the way of the mass of food. The passage from

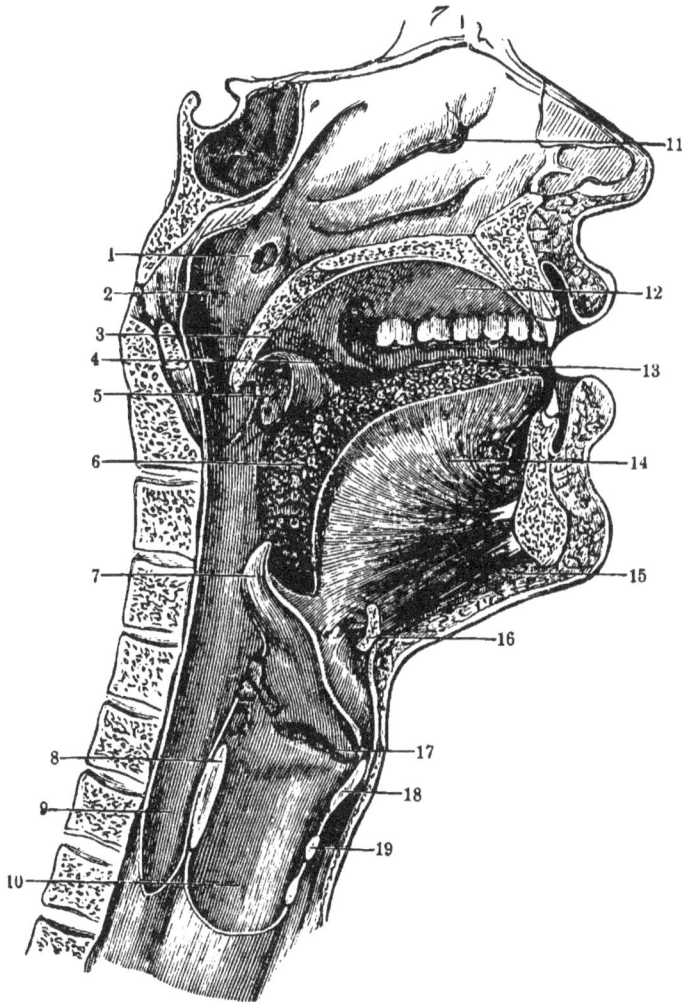

FIG. 24.—GENERAL VIEW OF MOUTH, PHARYNX, ETC.

1, canal from throat to middle ear; 2, back part of nose; 3, soft palate; 4, soft palate covering tonsil; 5, tonsil; 6, base of tongue; 7, epiglottis; 8, part of cartilage of larynx; 9, laryngeal portion of pharynx; 10, cavity of larynx; 11, nasal fossæ; 12, vault of the palate, or roof of mouth; 13, 14, tongue; 15, muscle beneath tongue; 16, hyoid bone; 17, interior of larynx; 18, 19, thyroid cartilage.

the back part of the throat into the nose is also closed by muscular contraction, so that fluids cannot get into that cavity. There is a cartilaginous body called the *epiglot'tis*, which is placed at the upper part of the larynx, behind the base of the tongue, and when the food is swallowed it covers the glottis, as the opening in the larynx is called, and thus aids in preventing the mass from getting into the air-passages. The passage into the stomach is effected first by the very rapid and convulsive contraction of the muscles of the throat or pharynx, and of other muscles which remove the larynx, or upper part of the air-tube, out of reach, and give a smooth surface for the food to pass over. In this way it reaches the œsophagus or gullet, which also contracts around it, forcing it into the stomach. The lower part of this tube remains contracted for a while after the entrance of the food into the stomach, so as to prevent its return upward.

Processes of Digestion.—The first step of digestion is called *prehen'sion* of the food, which really denotes nothing more than taking the solid or liquid food into the mouth. It therefore requires no elaborate description. Digestion in the mouth is called *o'ral* or *buc'cal* digestion (*os, oris,* or *bucca,* "a mouth"), including the action of the salivary glands, the teeth, etc., already described. Deglutition, or swallowing, is the next step of the process, after which come, in immediate succession, stomachal and intestinal digestion.

When the food is hastily swallowed—and this is generally called "bolting the food"—the cutting and crushing action of the teeth is not effected, and the alimentary substance is swallowed in a state unfitted for perfect digestion in the stomach, so that dyspepsia or indigestion may result, especially if this neglect becomes habitual. Some of the salivary glands pour out a thinner fluid than the others—the thicker secretion facilitating the progress of the alimentary mass, or bolus, as it is called, when swallowed, while the thinner helps to dissolve it. The important chemical change which takes place in the mouth, however, is the conversion of starchy matters in the food into dextrine and grape-sugar. This is effected by means of a ferment called *ptyalin* (pron. *ty'alin*, from a Greek word meaning saliva), which is present in that fluid. If the food escapes this action through hasty swallowing or other cause, another fluid which is poured into the intestines, called the pancreatic juice, or juice of the pancreas, exerts a similar effect upon it after it has passed out of the stomach. No detention of thin fluids takes place in the mouth ; they are not acted upon in that cavity, but in the stomach. In the case of drinking, fluids are usually sucked into the mouth and swallowed without any action upon them in that cavity. When the lips are applied to a cup the air is drawn inward by inspiring or breathing in, and the liquid flows into it.

As ptyalin is the ferment of the mouth, so do we have also a similar ferment in the secretion of the stomach, called *pep'sin*, but endowed with different powers; and in the pancreatic juice, *pan'creatin*, which exerts the effect already alluded to on starchy matters.

The voluntary and involuntary steps in the process of deglutition to which reference has been made show that there must be two separate divisions of the nervous system regulating it. This we find to be the case. When we come to consider the general subject of the Nervous System we shall find that it has its divisions into a voluntary and an involuntary nervous apparatus, the former alone being under the control of the brain. The moment the food

8

passes over the top of the windpipe to the œsophagus, the involuntary system comes into play. Otherwise respiration would be seriously interfered with, and danger to life result. The process of deglutition also includes the passage of the food along the œsophagus.

Digestion in the Stomach and Intestines.—When the food reaches the stomach it is subjected to entirely new processes,

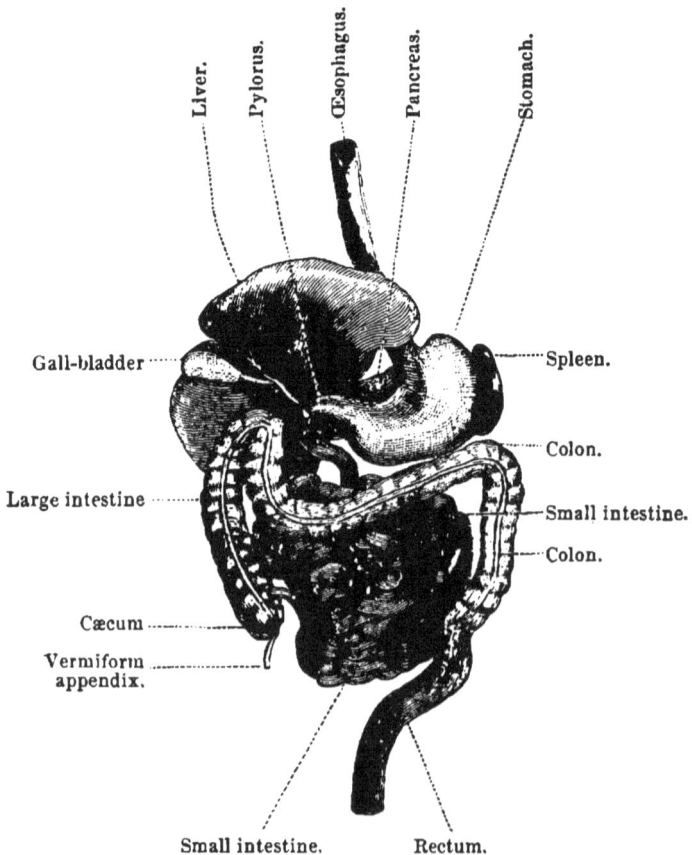

Liver.   Pylorus.   Œsophagus.   Pancreas.   Stomach.

Gall-bladder ···· ···· Spleen.

···· Colon.

Large intestine ···· ···· Small intestine.

···· Colon.

Cæcum ····

Vermiform ····
appendix.

Small intestine.   Rectum.

FIG. 25.—ALIMENTARY CANAL.

both chemical and mechanical. Here it remains for a greater or less time, according to its digestibility. The stomach

and intestinal canal of man occupy a medium place between the carnivorous and herbivorous animal, resembling
the latter rather than the former. In the carnivorous the
animal food is easily digested, and the alimentary canal is
not, therefore, long, as the food requires a very short stay
in the stomach and in those parts
of the intestines concerned in
active absorption. Herbivorous
animals require a great deal of
digestion of the food before it
becomes fit to be absorbed, and
the canal is therefore long, and
some parts of the intestine are
very capacious. The stomach is
also large in these animals and
complicated in structure, so as to
receive and digest food that is
bulky without being greatly nutritious. Some of these animals
have as many as four stomachs.

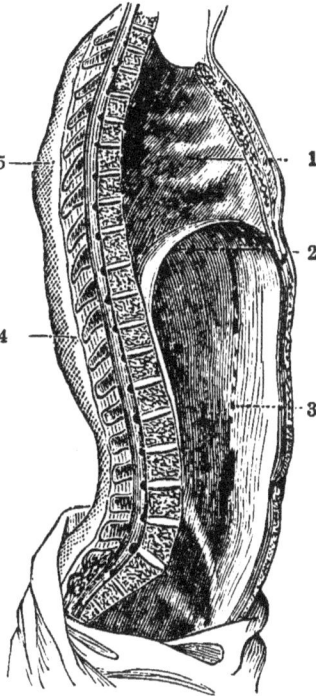

In man the stomach is the
most dilated part of the digestive apparatus, and is in shape
like the ordinary bagpipe. It
lies across the upper part of the
abdomen, and is separated from
the chest, in which the heart and
lungs are placed, by a thick muscle, called the *di'aphragm*
(Fig. 26), which covers the whole width of the floor of
the chest. The opening in the stomach at its left end
is for the entrance of the œsophagus. It is opened and
closed by a muscular arrangement. The opening at the
right side is the point at which the food passes into the

FIG. 26.—THE THORAX AND
ABDOMEN.

1, thorax, or cavity for heart and
lungs; 2, diaphragm; 3, abdomen;
4, spinal column; 5, spinal canal.

intestine, which here begins its course. The left end of the stomach is called the *card'iac extremity*, because it is near the heart, or *splenic*, because it is near the spleen; the other, the *pylo'ric* (from *pylorus*, " a janitor"), because of the existence of a kind of valve here which, like a janitor, will not ordinarily allow the food to pass into the intestine until properly prepared to do so. The stomach is lined by a

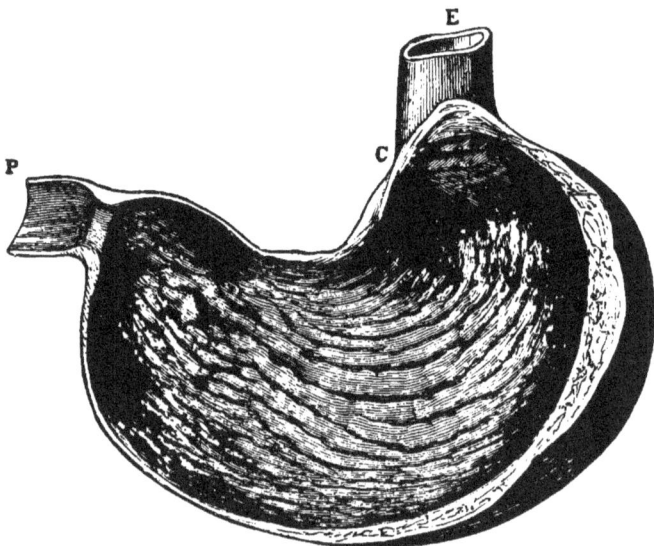

FIG. 27.—INTERIOR OF THE STOMACH.
P, pylorus; E, œsophagus; C, cardiac orifice of the stomach.

membrane (Fig. 27) secreting a fluid called mucus, and hence called a mucous membrane. The mucous membrane is not smooth, but is thrown into numerous folds. It is continuous from the inner edge of the lips through the whole alimentary canal. This mucus keeps the membrane always moist, and is the only fluid secreted when the stomach is empty.

When any article passes into that organ, however, a peculiar colorless fluid is poured into the interior of the stomach, called the *gastric juice*. It is chiefly water, but contains an acid—by some considered lactic acid, by others

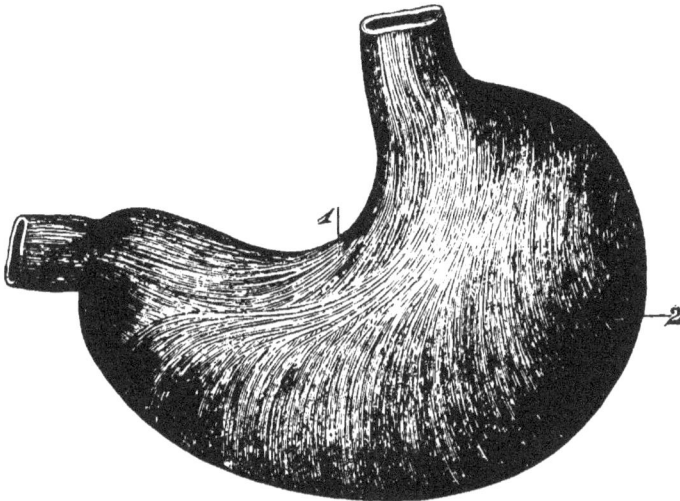

FIG. 28.—MUSCULAR FIBRES OF THE STOMACH.
1, circular fibres; 2, oblique fibres.

hydrochloric—and a ferment, to which the name of *pepsin* has been given. The internal surface of the stomach presents a network of minute ridges, in the spaces between which the mouths of little glands open. The entire structure is permeated with minute vessels, which enter these ridges and send numerous branches around these glands. Many little glands of different sizes and shapes are found in all

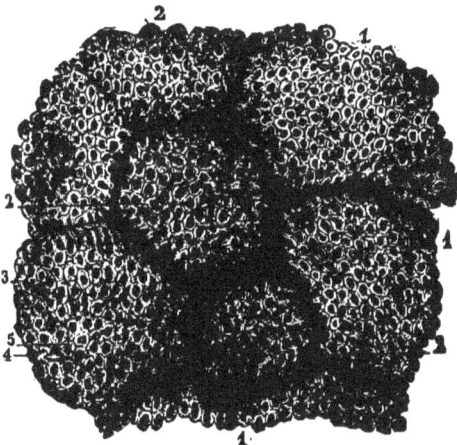

FIG. 29.
1, 2, 3, pits in mucous membrane of stomach;
4, 5, orifices of the glands.

parts of the stomach, but especially at the upper part.

8 *

They pour out a thin kind of fluid, and contribute their share to the formation of the juices that assist in digestion by the moisture they create. This fluid is always present, but the gastric juice, so called, is probably secreted more profusely in that part of the stomach lying in close vicinity to the intestine, and only after food is taken. The glands secreting the gastric juice are tubular in character (Fig. 30) and lined with cells. They are found more at the lower part of the organ.

FIG. 30.—GLAND FROM PYLORIC PORTION OF STOMACH (magnified). 1, duct or canal; 2, principal branches; 3, terminal portion.

The muscular coat of the stomach (Fig. 28) is outside the mucous one, but in contact with it. The contractions of the stomach, which produce the churning movement so necessary to perfect digestion, are due to the presence of the muscular coat, which is made up of fibres running in different directions. Some are longitudinal, some oblique, others circular. The effect is to contract it in every direction. Should all these fibres contract at the same time, the stomach would be forcibly emptied, the food being expelled from it. The longitudinal fibres, acting alone, contract or shorten the stomach. The circular fibres, especially at the pylorus, press on the contents of the stomach, and cause the food to pass from one end to the other. The bloodvessels and nerves of the stomach are more numerous than those of any other organ of the body. The effect of the contraction of the oblique fibres in changing the shape of the stomach is well exhibited in Fig. 31, the organ being divided into two distinct portions. The outer

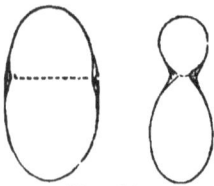

FIG. 31.

coat of the stomach—the serous—is a delicate structure, not concerned in stomachal digestion.

Stomachs of other Animals.—In man the stomach is not

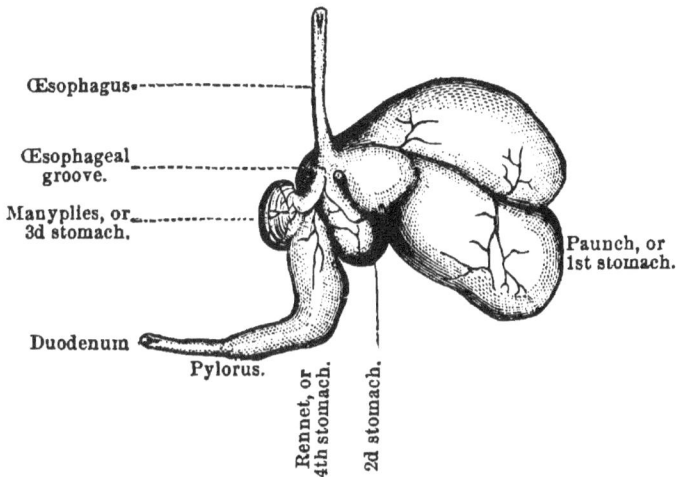

Œsophagus
Œsophageal groove.
Manyplies, or 3d stomach.
Paunch, or 1st stomach.
Duodenum
Pylorus.
Rennet, or 4th stomach.
2d stomach.

FIG. 32.—STOMACHS OF A SHEEP.

as small as in the carnivorous animal. In the herbivorous the stomach is very complicated. In the ox, for example,

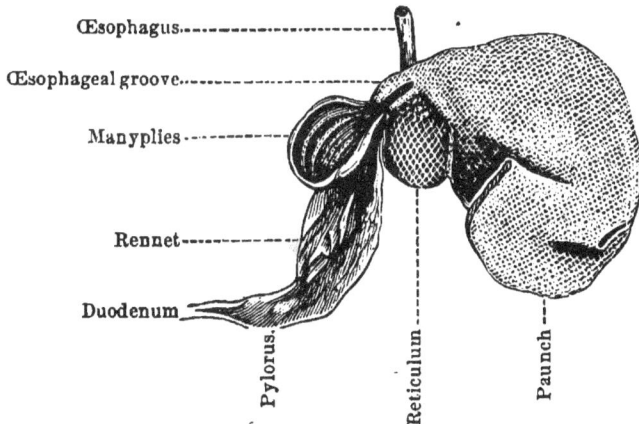

Œsophagus
Œsophageal groove
Manyplies
Rennet
Duodenum
Pylorus.
Reticulum
Paunch

FIG. 33.—STOMACHS OF A SHEEP (INTERIOR).

there are four distinct divisions, all of which are interested in the digestive process, but of which the fourth is the only one resembling the human stomach. Some of these animals have the power of returning the food from the second stomach to the mouth to be *ru'minated*, as it is

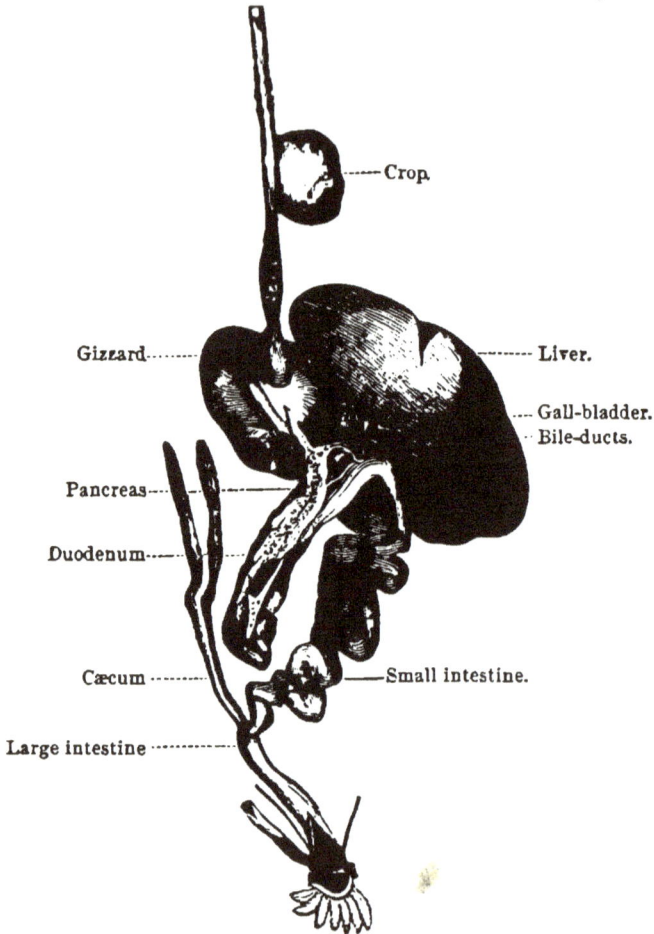

FIG. 34.—DIGESTIVE APPARATUS OF DOMESTIC FOWL.

called, or chewed over again ; and when it comes down again it passes directly into the third stomach without entering the first and second. The peculiar arrangement of a sheep's stomach is shown in Figs. 32, 33. In *birds*, in which grain or seed is swallowed whole, and therefore unfit for digestion, a sac called a crop (Fig. 34) exists, in which fluid is poured out to aid in its digestion. There is also another small cavity or second stomach, which leads into a third stomach or gizzard, a kind of paunch connected with the alimentary canal proper. This organ is a powerful muscular arrangement, a thick sac with a small cavity containing a number of pebbles, which seem to take the place of teeth, and by their mill-stone action crush the food by trituration or grinding. In birds of prey the crop is less developed. In carnivorous or flesh-eating birds the gizzard is not so powerful an organ as in grain-eating birds, and the alimentary canal is altogether much more simple, as the food does not require so much digestion. In some of them the stomach is merely a sac of muscles and membranes, and pours out a powerful secretion to act upon the food. This is all the more necessary as the animal does not possess any arrangement for mastication, and it is essential that the digestive organs shall be packed away in a small space, so that its flight will not be interfered with.

In *reptiles*, a class of animals which is capable of abstaining from food for a long time, there is generally a mouth of large capacity, with teeth that form a part of the jaw itself, or, like tortoises, their jaws are horny; the stomach varies in size and shape, and the intestines are usually short.

In *fishes*, those which feed on the blood and juices sucked from other animals have their mouths, teeth, and tongue arranged and acting upon the principle of the cupping apparatus of the surgeon. In all the liver is large, the stomach and intestines varying in size, and digestion being performed quite rapidly.

Insects (Fig. 35) vary in the arrangement of their digestive apparatus, as in their nutriment. Some feed on

the juices, others on the solid parts, of both plants and
animals. Where it is necessary to cut the food, as in the
case of the grasshopper, a cutting arrangement exists, in
addition to feelers, a tongue, etc. In herbivorous insects,
the digestive apparatus is considerably developed and complicated. There are really three stomachs—a crop, gizzard,

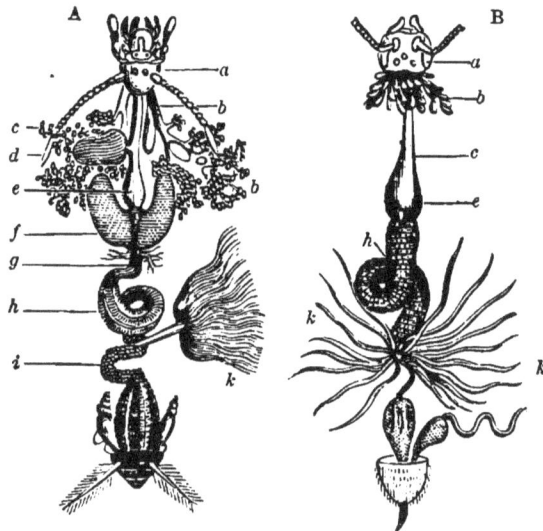

FIG. 35.—DIGESTIVE APPARATUS OF INSECTS.

A, MOLE CRICKET.—a, head and appendages; b, salivary glands; c, secreting granules
of the same glands; d, feelers; e, cardiac part of stomach; f, accessory pouches
of the stomach; g, middle part of stomach; h, pyloric part of stomach; i, intestines; k, canals representing liver.
B, BEE.—a, head and mouth; b, salivary glands; c, œsophagus; e, crop; h, stomach;
k, canals representing liver.

and stomach proper. The alimentary canal, including the
intestines, is not straight, but generally tortuous, with enlargements and contractions at various points.   In those
which live chiefly on animal food the canal is short, and
in those which feed on vegetable substances the canal is
longer. This we have already seen to be true of animals
much higher in the scale.

In all forms of animal life there is a stomach in some shape or other, although it may be, in its simplest expression, nothing more than a mere expansion or sac, through which all digestive and nutritive absorption is effected. In some of the very lowest animals, indeed, as already stated, an intestinal canal is wholly absent, the expansion referred to taking the place and performing the duties of both stomach and intestines.

The variety of forms of stomach characteristic of different types of animals is well shown in Fig. 36:

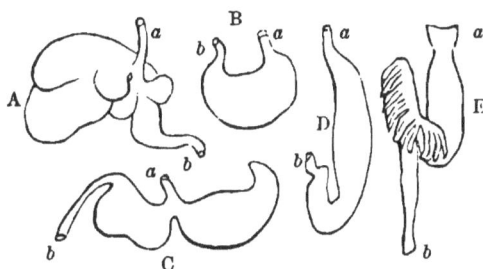

FIG. 36.—STOMACHS OF VARIOUS ANIMALS.

A, sheep; B, hyena; C, marmot; D, seal; E, salmon. a, cardiac opening of stomach; b, beginning of intestine.

Action of the Stomach in Digestion.—When the food enters the stomach to be digested, that organ gradually becomes distended, especially in its muscular coat, so as to occupy a much larger space in the abdomen. The dilatation takes place chiefly forward and upward and to the left, so that the stomach experiences a movement of rotation. When it becomes overloaded it may press upward against the floor of the chest, so as to make respiration laborious. A regular series of contractions and relaxations of the stomach takes place under the exciting presence of food—regularly and alternately from left to right, from the opening of entrance to that of departure, and then in the opposite direction. The result is the admixture of the food with the fluids secreted in the stomach, and the bringing of fresh portions under their influence. From the time of entrance of food into the stomach at the cardiac extremity until its appearance at the pylorus, when ready to make its exit into the intestine, the contents of the stomach are undergoing a regular revolution or churning process, each revolution taking perhaps from one to three minutes in its accomplishment. The diaphragm, the large muscle separating the chest from the abdomen (Figs. 26, 49), by its powerful contractions also presses upon the stomach and increases its action. The movement of the stomach and also of the intestines has been considered so worm-like in its character as well to deserve the name *vermic'ular* (*vermes*, "a worm") assigned to it. Stomachal digestion may therefore be said to include, as its two important elements, this gentle motion of the organ and admixture with its fluids.

Changes in the Food in Digestion.—The name *chyme* (pron. *kime*) has been given to the soft, pulpy, semi-fluid mass undergoing digestion in the stomach. It is no longer the food proper, but food combined with secretions from the salivary glands and the stomach. It of course differs in character with the nature of the food taken, the amount of liquid which may have been swallowed, and with the stage of the digestive process at which it is examined. Saliva is probably poured into the stomach from the mouth during the whole of stomachal digestion. If the food taken has been starchy, it is probable that some of it has undergone conversion in the mouth, as previously stated; but some of it that was unconverted will be acted upon by the saliva in the stomach, or may pass into the intestine without further change. Fatty matters also wait there unchanged, to be digested after they leave the stomach.

Other materials, such as the albuminous, fibrinous, etc.,

of which meat is composed, undergo their digestion in the stomach itself. Besides all these there will be found a variety of indigestible matters which are of no use to the system, and must eventually be thrown out of the canal. In the decomposition of the food acids will be formed which have no part in the digestive process, such as acetic acid from fermentation of the sugar, etc. All kinds of food are not, therefore, wholly digested in the stomach. Oily matters undergo digestion in the intestine, and as a rule vegetable food also awaits its exit from the stomach to seek the action of the intestine. Thick fluids, such as soups, may be acted upon in both regions.

Animal food is more digestible than vegetable on account of the peculiar arrangement of its fibres. Animal fats and other oils are not only indigestible in the stomach, but they interfere with the process of digestion there, and are not themselves easily digested when they pass into the intestine. Bread, potatoes, and pastry are partly digested in the stomach and partly in the intestine. Liquid substances, as will be shown hereafter, are generally absorbed by the bloodvessels of the stomach at once if sufficiently thin.

The *length of time* required for digestion of the food in the stomach, under favorable circumstances, is two to four hours, but this period is lengthened in those who are of languid and sedentary habits, and in those who do not take the proper amount of exercise. While the food is in the stomach it is exposed to a temperature of at least 100° Fahr. Under the influence of all these agencies—gastric juice, acids, heat, etc.—the animal food is converted into a modified form of albumen, called pep'tone or albu'minose, which is capable of more ready absorption than albumen. The active principle of the gastric juice—pepsin, obtained from it by chemical action—is the important medium

9   G

through which these changes are effected. The gastric
juice is a powerful antiseptic; that is, it will check the
putrefaction of substances like meat. The main result,
therefore, of stomachal digestion is the transformation of
alimentary matter into chyme, and the change effected upon
animal food. The action is both chemical and mechanical.

Action on Thin Fluids.—Certain thin fluids furnish
neither chyme nor chyle, being essentially composed of
water, or water mixed with alcohol or other materials.
Although really a variety of Absorption, which we shall
soon consider, we may briefly allude to this subject under
the head of Digestion. These enter the stomach without
being subjected to any action in the mouth, and are absorbed by the veins of the stomach without undergoing
any other action, and by the veins of the small intestine.
The veins into which they enter unite with other veins to
form what is called the *por'tal vein*, which goes directly to
the liver. After a short stay there this blood passes into
the current of the blood to the heart and through the system
generally. Not being digested, but passing rapidly into
the circulation, we can understand why the effect of alcohol
is so soon felt when taken into the stomach, and why the
liver should become diseased in those who drink liquor to
excess. Other nutritious substances, if in very fine solution, also probably pass into the same veins without being
subjected to the process of digestion.

Digestibility of Food.—Substances are said to be very
digestible or easy of digestion if they are known to make
but a short stay in the stomach; but some of these very
articles may not be nutritious, and may be difficult of
digestion in some healthy but peculiarly constituted
stomachs. The terms *nutritious* and *digestible* do not,
therefore, imply the same thing. It seems to be the rule

that substances which are capable of affording nourishment are detained for a longer time in the stomach than others, as if that organ were endowed with a power or sense of choosing what is best for it. Some materials, as crude vegetables, oils, and fats, will soon pass out of the stomach without being digested there at all. It has been contended by some that tables showing digestibility of various articles of food seem to exhibit only how long each remains in the stomach, but they do not prove the exact degree of digestibility. Milk probaby disappears from the stomach very rapidly. It is said by one writer that an hour after it is swallowed there are scarcely any traces of it in the stomach; but milk is composed not only of sugar, water, and salts, all of which may be absorbed in that time, but it also contains fatty matter, which is not so easily gotten rid of. It is estimated that five or six pounds of gastric juice are secreted daily; ten or twelve pounds, according to some.

Aids to Digestion.—Proper digestion is rendered more probable by rest of mind, agreeable mental and gentle physical exercise, as well as by variety of food. Over-exertion of any kind should be avoided soon after taking a full meal. The recumbent posture should not be indulged at such times, and prolonged sleep should be deferred until after digestion has been accomplished. If indulged in at all, a short nap may be taken in a sitting posture. Digestion in the earliest periods of life may be impaired by the tendency to excess of acid in the stomach, but usually the digestive powers of young persons are better than those of adults, considering the food which they are each capable of taking. Where articles disagree, the fashion has often been to say that the individual is *bil'ious;* but such a term conveys no idea whether there is too much or too little bile poured into the intestine for digestive purposes. Of course under irritation of food that is indigestible the liver may be excited to renewed energy, and the action of the muscles of the upper part of the intestine may be inverted and bile be brought into the stomach.

In regard to taking *fluids at meals,* if they are not stimulants the habit should be regulated by the inclination. With some persons little or no drink at such times is necessary, and drink can be taken between meals. It is certainly an error of diet to wash down every portion of solid food with liquids, as is the habit of some. A moderate quantity of liquid is generally desirable, and after the gastric juice has ceased to act it may be excited to action by a fresh supply of water. It is thought by many conducive to health to take a glass of pure water in the morning before breakfast. Exercise before breakfast disagrees with many a weak stomach and produces a languid feeling all the day. Long walks before the morning meal are not attended with beneficial results as often as the devotees of systematic training claim for them.

Length of Time required.—The following table gives the result of experiments to determine the length of time which some of the more commonly employed articles of food occupy before their disappearance from the stomach. This represents, to some extent, their digestibility, and also the effect of cooking upon the substances named:

	HOURS.		HOURS.
Apples, raw	2	Butter, melted	3.30
Barley, boiled	2	Cabbage, raw	2.30
Beef, roasted	3	" with vinegar	2
Beefsteak, broiled	3	" boiled	4.30
Beef, fried	4	Cake, corn, baked	3
Beets, boiled	3.45	" sponge, baked	2.30
Brains, animal, boiled	1.45	Catfish, fried	3.30
Bread, corn, baked	3.15	Cheese, old, strong	3.30
" wheat, baked	3.30	Chicken, fricasseed	2.45

	HOURS.		HOURS
Corn and beans, boiled	3.45	Pork, roasted.	5.15
Custard, baked	2.45	" salted and fried	4.15
Duck, roasted	4	Potatoes, Irish, boiled	3.30
Dumpling, apple, boiled	3	" roasted	2.30
Eggs, hard boiled	3.30	Rice, boiled	1
" soft boiled	3	Salmon, salted	4
" fried	3.30	Soup, barley, boiled	1.30
" roasted	2.15	" bean	3.30
" raw	2	" chicken	3
Fowls, boiled	4	" mutton	3.30
" roasted	4	" oyster	3.30
Goose, roasted	2.30	Tapioca, boiled	2
Lamb, boiled	2.30	Tripe, soused	1
Milk, boiled	2	Trout, salmon, boiled	1.30
" raw	2.15	" " fried	1.30
Mutton, roasted	3.15	Turkey, roast	2.30
" . broiled	3	" boiled	2.20
" boiled	3	Turnips, boiled	3.30
Oysters, raw	2.55	Veal, broiled	4
" roasted	3.15	" fried	4.30
" stewed	3.30	Vegetables and meat, hashed	2.30
Pig, roasted	2.30	Venison steak	1.35
Pigs' feet, soused	1		

Digestion in the Intestines.—The intestines are a continuous canal, leading directly from the stomach. They vary in length in different animals, and are usually divided into the small and the large intestines. The small intestine (Fig. 37) is a narrow flexible tube about twenty feet long, coiled upon itself, and divided into three parts—the *duode'num*, so called from its length being estimated at twelve fingers' breadths (*duodeni*, "twelve"); the *jeju'num* (*jejunus*, "empty"), so called from its usually being found empty after death; and the *il'eum* (from a Greek word meaning "to twist," on account of its convolutions). It is in these portions of the intestine that intestinal digestion chiefly takes place. The large intestine, which is intended rather

9 *

as a receptacle for the useless and undigested portions of
the food, is divided into the *cœ'cum* or blind sac (*cœcus*,
" blind "), the *co'lon*, and the *rec'tum* (*rectus*, " straight ").
The intestines have similar coats to the other parts of the
canal, being lined with mucous membrane, which is covered
by a muscular coat, and this again by a serous coat, here

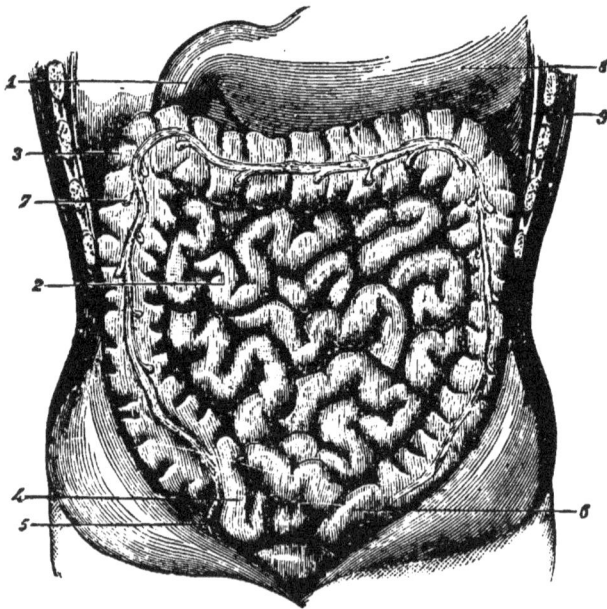

FIG. 37.—GENERAL VIEW OF THE INTESTINES.
1, duodenum; 2, small intestine; 3, large intestine; 4, ileum; 5, appendix;
6, rectum; 7, colon; 8, stomach; 9, diaphragm (cut).

called the *peritone'um*. The same vermicular or worm-like
movement of alternate contraction and dilatation takes place
here as in the stomach. This is beyond the control of the
will, and is excited by the presence of food. The circular
muscular fibres contract the intestine in its calibre, and the
longitudinal fibres shorten it in its length.

The arrangement of the intestines in the lower animals
is somewhat different. In man there is a small pouch con-

nected with the cæcum, called the *appendix*, which is of no
practical use whatever, but it seems to be left there to show
a similarity in original structure in the animal series. In

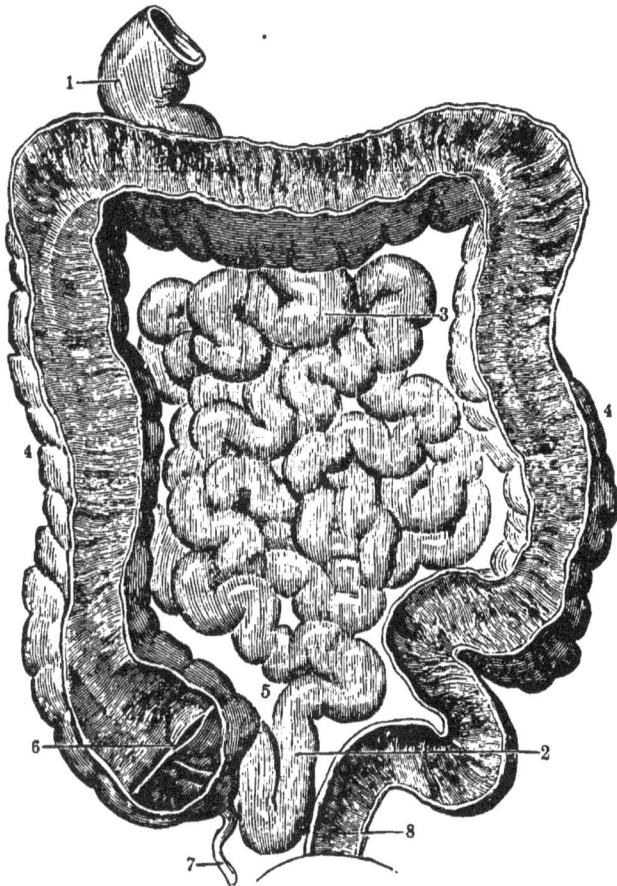

FIG. 38.—THE LARGE INTESTINE (OPENED).
1, duodenum; 2, ileum, connecting with small intestine; 3, jejunum; 4, 4, 4, colon;
5, cæcum; 6, valve between ileum and cæcum; 7, appendix; 8, rectum.

some animals it is so large as to be really a distinct intestine. In birds that eat grain or that feed on all kinds of
food there are two cæcums. In reptiles the intestines are
short, and there is but little difference in size between the

large and small intestines.   In insects the intestinal canal is
often considerably developed and more complicated than in
some of the upper classes of animals.
In fishes the pancreas is absent, but
is sometimes replaced by cæcums
around the pylorus (Fig. 39).

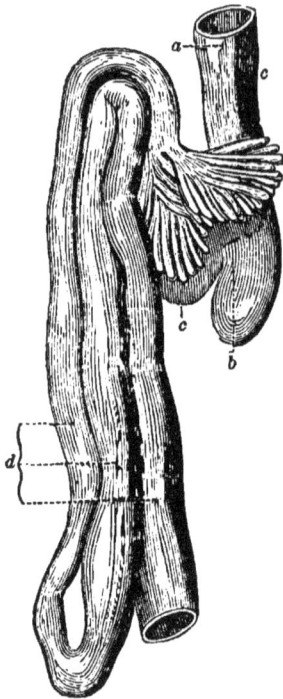

Intestinal Juices.—In man the
mucous membrane of the intestines
is thrown into permanent folds, so
as to give a large surface for absorption, and bloodvessels, absorbent vessels, and nerves are largely
distributed in them.   Here, too, on
the mucous membrane is poured out
a fluid called the *intes'tinal juice*,
which is secreted by an immense
number of little bodies or glands.
It has been estimated that there are
many thousands of these little secreting bodies, or fol'licles as they
are called, engaged in this duty in
a single individual.   Two anatomists, named Brunner and Lieberkühn, first described these little
glands, which are now called, after them, the glands of
Brunner and Lieberkühn.   This is not the only fluid poured
into the intestine, for at a distance of four or five fingers'
breadths from the pylorus there open into the duodenum
two canals, which bring *bile* from the liver and gall-bladder,
and a fluid called the *pancreatic juice* from an organ called
the pancreas, or abdominal sweetbread, as it is sometimes
termed, lying close to the stomach.   Both of these fluids aid in
the digestion of the food in that portion of the intestine.   The

FIG. 39.—INTESTINE OF BONY
FISH (MACKEREL).

a, œsophagus; *b*, stomach; *c*, pyloric cæcums; *d*, intestines.

arrangement of the pancreas
and the opening of its canal
into the intestine are shown
in Fig. 41.

It has been estimated that
more than twenty pounds of
digestive fluids are secreted in
the twenty-four hours at different parts of the apparatus
for digestion :

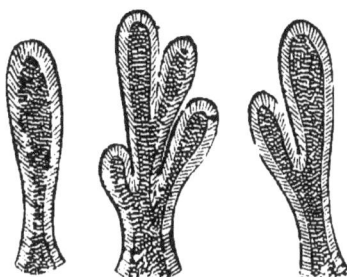

FIG. 40.—TUBULAR GLANDS OF IN-
TESTINE (magnified 100 times).

	Pounds.		Pounds.
Saliva	3½	Pancreatic juice	½
Bile	3½	Intestinal juice	½
Gastric juice	12		

It seems worthy of remark that the food is acted upon
in the mouth by an alkaline material; in the stomach by

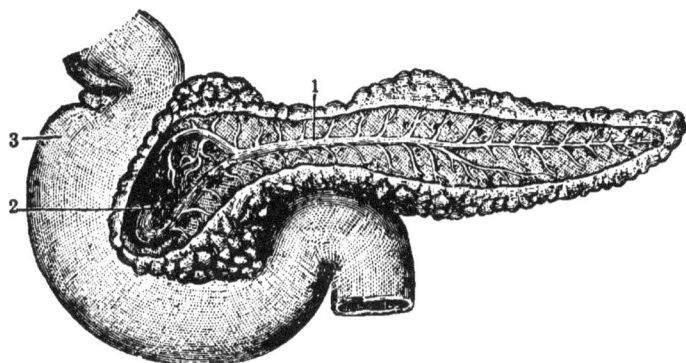

FIG. 41.—THE PANCREAS,

Divided, so as to show—1, pancreatic duct or canal; 2, canal emptying into
intestine; 3, duodenum.

an acid; in the pancreatic fluid again by an alkaline fluid.
This alternate chemical action is doubtless connected with
the perfecting of the digestive process, although not easy
to explain.

When the valve between the stomach and intestines

allows the food to pass, after digestion in the stomach has been effected, the muscles of the latter organ drive it out into the duodenum or first part of the intestine. This valve—the *pylo'rus*—is not at all under the control of the will, and acts wholly under the stimulus of the food.

In the duodenum the food is mixed with the bile and pancreatic juice, and with the intestinal juice already referred to. The effect of the pancreatic fluid on the food is to make a mixture, resembling what the apothecary calls an emulsion, by the admixture of fatty or oily matters with other ingredients. When this emulsion is made by the action of the pancreatic juice on oily or fatty food that has escaped digestion in the stomach, it is more easily absorbed for the useful purposes it has to subserve in the system. Besides this, the starchy parts of the food that were not converted into grape-sugar in the mouth undergo this change when brought into contact with the pancreatic juice. This fluid also has some action on albuminous matters akin to that exerted by the gastric juice.

The *bile* is secreted by the liver, an organ lying under the diaphragm, above the stomach. This is a very large organ, weighing three or four pounds, and measuring about ten or twelve inches in width. Only a small part of the bile formed by it and poured into the intestine is of use in digestion; the rest of it passes along with the refuse food. Bile is one of the chief agents in the digestion of fats. There is a sac connected with the liver called the *gall-bladder* (Fig. 25), and this acts as a receptacle for the bile. The canal from this and the one from the liver unite, so as to pour the bile from each together into the intestine. The gall-bladder is indeed filled from the duct or canal from the liver before it discharges itself, drop by drop, into the

canal common to the two. While food is undergoing digestion in the stomach the bile from the gall-bladder, as if by
a telegraphic signal from the stomach, begins to come down
into the intestine, to be ready for its duty when the chyme
or digestive mass arrives. This, for want of a better term,
is said to take place by *sympathy*, and the fact has been
proved, like many others here detailed, by experiments
made, in the interests of science, upon living animals. It
seems remarkable that so large a quantity of bile should
be poured into the intestine when we can ascribe to it so
little efficacy in the digestive process. It seems as if the
bile also acts as a stimulant on the mucous and muscular
coats of the intestines.

Action of the Intestines.—The vermicular action of the
intestines, sometimes called the *peristal'tic* action, passes the
food slowly onward from one portion to another. The
mixture of intestinal fluids with food forms a milky fluid,
called *chyle* (pron. *kile*, from a Greek word meaning "juice").
The process by which it is formed is called *chyl'ification*.
As has been stated, the mucous membrane of the intestine
is thrown into folds, and numerous small projections occur
in these (Fig. 42), called *villi*. In these commence a series
of vessels, called *chylif'erous vessels*, because they carry away
the chyle, or *lac'teals*, because the fluid they contain is milky
in appearance (*lac*, "milk"). As will be described under
the head of Absorption, the chyle is absorbed into these
vessels and carried directly into the blood. The intestinal
juice proper, it is thought, has very little effect other than
that of moistening the food. It is not acid, like the gastric
juice, and being alkaline it gradually neutralizes the acidity
of the latter, and thus prevents fermentation and decomposition of the food. By the time that the food has reached
the large intestine all the nutritive matter has been taken

from it. Digestion proper may be said, therefore, to be virtually ended with the small intestine.

A certain amount of air is always present in the intestines, dependent chiefly on the chemical changes in the food during digestion. The air existing in the stomach contains oxygen, but this chemical element is not usually found in the intestines.

With this sketch of the process of digestion the reader can readily appreciate the manner in which nourishment is disposed of to meet the wants of the body. Nearly all the food taken into the mouth is digested in one form or another. To sum up the results at which we have arrived, we may briefly say that when solid food is taken it is disposed of as follows: the starchy matters are converted in the mouth and stomach, by the action of the saliva, into grape-sugar, and also by the pancreatic juice in the small intestine; fatty matters are chiefly digested in the small intestine by the pancreatic juice and bile together, being made into an emulsion for easy absorption by the chyliferous vessels; albuminous matters, such as meat, are acted upon by the gastric fluids.

QUESTIONS.

What is digestion?

In what part of the body is it effected?

Why does food require to be digested?

How is the general system affected by such a process?

What general action on the food takes place during digestion?

Through what circulating fluid is the body nourished?

What is the process called by which the digested food is carried into the blood?

Is digestion simply confined to animal life?

How is food supplied to the vegetable?

What chemical element of the plant is derived from the soil?

How can the soil be rendered more porous?

In what way is the plant nourished?

From what source is the food of man derived?

What is an aliment?

What is chyle?

Are any other substances used as food except those derived from the animal and vegetable kingdoms?

How is the growth of the body effected in early life?

Why does growth remain stationary in adult life?

On general principles, what ingredients are best adapted for food?

What simple article may be mentioned as an example?

What special articles of diet resemble the blood in composition?

What ingredients do they possess in common?

Through what fluid is the system largely supplied with mineral matters?

What substances are included under condiments?

What effect has salt upon the process of digestion?

In what part of the body does lime exist?

What are the principal chemical substances found in the body?

Which of these is most important as a constituent of the blood?

What is the average amount of iron in the human body?

Of what two chemical substances is starch composed?

What other articles of diet have the same composition?

What articles of food are mainly composed of starch?

What articles furnish cane-sugar?

What substances furnish grape-sugar?

In what organs of the body do we find sugar, and what kind?

In what form are fatty substances taken as food?

What are nitrogenized aliments? Albuminous?

From what part of mineral food are they derived?

Are they found in vegetables?

What is the effect of acid drinks in digestion?

Into what two classes has food been divided?

Which predominates in animals?

Why is nitrogen important as an element of food?

Why does a horse consume more food, in proportion to its weight, than a dog?

What is the effect on animals of restriction to one class of food?

Why can they thrive on milk alone?

What disease arises from want of variety in diet?

On what principle are diet-tables for the army and navy founded?

10

From what are oils and fats derived?

What three chemical substances are found in them?

What is the sweet principle of fats called?

What is the difference between an oil and a fat?

What is the appearance of oily matter under the microscope?

What is an emulsion?

What important fluid is an example of an emulsion?

In what part of the world are oily matters taken largely as articles of food? Why?

From what sources is sugar derived?

In what form do vegetables and fruits present it?

What are the digestive qualities of sugar?

What varieties of sugar are usually employed?

What is the purest form of starch?

With what other substance is it mixed in oats? Beans? Peas? Potatoes? Almonds?

From what substances is starch prepared artificially?

Is starch soluble?

What is said of gum as an article of food?

What are the three classes of albuminous substances?

From what does albumen derive its name?

In what substances is it found?

What is the chief ingredient in a bird's egg?

In what part of the body do we find fibrin in a fluid state?

What is gluten?

How is it obtained from wheaten flour?

To what does bread owe its porous character?

What are the principal ingredients of gluten?

In what important fluid do we find casein?

What is curds and whey?

What is rennet?

What chemical substance is specially characteristic of albuminous articles?

With what other substances should they be combined in a properly-mixed diet?

State from the table the main ingredients in bread, rice, milk, flour.

What is gelatin, and how is it obtained?

What is the effect of restricting an animal to one kind of food?

What portion of animals is generally employed for food?

Why was blood prohibited as an article of diet for the Hebrews?

What is the objection to the use of pork as an article of food?

What circumstances affect the character of animal food?

What element exists to a marked extent in the flesh of young animals?

What difference is there in young and older animals as to the relations of fat and flesh?

How is the flesh of animals affected by season?

Why are fat animals more desirable as sources of food?

What effect has the juice of the stomach on putrid substances?

What parts of birds are used as food?

What effect has exercise on the quality of the flesh of birds?

What are the principal ingredients of the flesh of fish?

What relation does it bear in its nutritive properties to a meat or a vegetable diet?

What term is applied to fish-eating tribes?

What chemical element in the flesh of fish renders that kind of food desirable in some conditions of the brain?

Why is milk well adapted to the nourishment of the body?

What change takes place in milk after it is taken into the stomach?

What becomes of the curd? Of the whey?

Why does the cheesy part of milk sometimes disagree?

How can milk be made more digestible?

What articles are derived from milk?

What are the constituents of cream, and why is it less digestible than milk?

What is butter, and why is it not very digestible?

What is cheese? Buttermilk?

Why is buttermilk more digestible?

What is koumyss?

Into what two parts is the egg divisible? Which has the most albumen? Which is most digestible?

What effect has heat upon the albumen?

What advantage has the egg as an article of diet?

Why are fried eggs less digestible?

What is an herbivorous animal?

How does an animal obtain nitrogen?

From what is bread usually made?

Of what is it chiefly composed?

What is leavened bread?

What is the effect of a ferment like yeast?

What changes take place in milk from fermentation? In bread?

What is said of the digestibility of cakes and pastry?

From what other flours is bread made?
How does bread aid the digestion of soup?
Why is stale bread more digestible than fresh?
What rule should be observed in regard to eating fruit?
How do preserved fruits act on the digestive organs?
On what does the value of food depend?
What effect has cooking upon food?
In what way does it render it more digestible?
What are the limits of temperature for cooked meats?
What is the effect of roasting on meat?
What is broiling, and what effect has it on meat?
What changes take place in meat when boiled?
What effect have baking and stewing?
Why are fried meats less digestible?
What effect has cooking upon vegetables?
What are the principal drinks used by man?
What rule should be observed as to the amount of drink to be taken
in the day or at meals?
What effect have drinks taken immediately before meals?
How do hot drinks affect the stomach?
What qualities should water possess for drinking purposes?
What does the drinking of water replace in the system?
What kinds of water are used for drinking purposes?
· What is the purest form of water?
What is the difference between hard water and soft water?
Which of these is preferable for cooking purposes?
Why is spring water more sparkling?
Why is well water less pure?
Why is lake water generally unfitted for drinking purposes?
How can foul water be rendered purer?
What is distilled water?
What is wine?
To produce a wine, what materials must be present?
How do red wines differ from white?
What produces effervescence in wines?
How are sweet wines made?
What are the chief ingredients of wine?
To which ingredient is its strength due?
What is the proportion of alcohol in wines? In spirits?
Why is old wine more desirable?
How are the alcohol and carbonic acid produced in wines?

How do malt liquors differ from wines?

What proportion of alcohol do they contain?

What is malt, and how are ales and beers derived from it?

What is the ferment, and how does it act?

How are spirits produced from wines?

What is brandy? Whiskey? Rum? Gin?

How do coffee, tea, cocoa, etc. resemble each other?

How do coffee and tea differ?

What is the difference in effect on the system of these and alcoholic drinks?

On what does the quantity of food to be taken depend?

What is the estimate of the quantity of meat, bread, fat, and water daily consumed by a person in full health?

What effect has temperature on the amount?

At what period of life is more food required?

What effect has exercise?

What rules should be observed as to meals?

Why are late or meat suppers objectionable?

How should the diet of those with weak digestive powers be regulated?

What processes are interfered with by hasty swallowing of the food?

What three conditions are necessary for a healthy diet?

How is the internal heat of the body kept up in cold climates?

What is the natural desire for food called? The artificial desire?

To what organ is the sensation referred?

When does it occur?

At what age is it most urgent?

What is the cause of hunger?

What is the effect of protracted hunger on the system generally?

Which of the functions give way last?

In what way does the brain suffer?

What action of the stomach has been wrongly assigned as the cause of hunger? Why does this not explain it?

What is the fluid poured into the stomach during digestion called?

How do we know that this is not the cause of hunger?

Is the expression of hunger restricted to the animal?

In what way does the plant exhibit it?

What is thirst?

What are the causes of the sensation?

Is thirst or hunger more imperative?

Why does salty food increase the thirst?

Can thirst be relieved by external means?

10 *   H

When does death usually result from protracted thirst?
What is the simplest form of digestive apparatus?
What is the simplest arrangement in the animal?
What do we next find as we rise a step higher in the animal scale?
What is the digestive arrangement in the infusory animalcule?
What is the general arrangement for digestion in man and the higher classes of animals?
Into what parts is it divided in man?
Which of the organs named occupy the abdomen?
To what action is the food subjected in the mouth?
What is the fluid poured into the mouth called?
What name is given to the organs producing the saliva?
What is insalivation?
What is the action of the teeth upon the food called?
How is their action increased in some animals?
What is the chief material in teeth?
What are the different parts of a tooth called?
What do we find in the interior of a tooth?
What are the earliest teeth called, and how many are they?
How do we divide them?
When does each kind make its appearance?
What change takes place in the teeth about six or seven years of age?
How many permanent teeth are there?
How are they arranged?
What are the wisdom-teeth?
Why are incisors so called?
Why are canine teeth so called?
What is the arrangement of the jaws and teeth in animals that live on flesh?
What information may we learn from a single tooth?
What is a carnivorous animal?
What term is applied to an animal that feeds on grain? To grass-eating animals?
What is an omnivorous animal? An insectivorous animal?
What is the arrangement of the teeth in the insectivorous animal?
What is a frugivorous animal, and what is the peculiarity of its teeth? Of the grain-eating animal?
What name is applied to the grinding teeth?
What animal has molar teeth almost exclusively?
What relation do the teeth of man bear to the herbivorous and carnivorous animal?

What action has the tongue in digestion?

What other action do the tongues of animals exert?

What is the arrangement in the elephant for taking food? In insects?

Where are the salivary glands situate?

What is their number, and what do they pour out?

What effect has the presence of food in the mouth upon these glands?

What is mouth-watering?

Is the quantity of saliva affected by the kind of food?

What changes take place in the food from the action of the saliva?

Is digestion in the stomach assisted by such action in the mouth?

What is the interior structure of the salivary glands as seen under the microscope?

How much saliva is secreted in the twenty-four hours?

What is meant by the terms parotid, submaxillary, and sublingual applied to the salivary glands?

What is the process of deglutition?

What organs are concerned in deglutition?

After the food leaves the mouth into what cavity does it pass?

What other cavities open into the pharynx?

What is the soft palate?

What canal leads from the pharynx to the stomach?

How long is it?

What kind of membrane lines it?

What part of the act of swallowing is under the control of the will? Which is involuntary?

Why is this last a wise provision of Nature?

How does the act of swallowing affect the movement of the larynx?

Why does not food pass into the cavity of the nose when swallowed?

What is the epiglottis, and how does it protect the larynx at such a time?

What action takes place just as food reaches the stomach?

How do the pharynx and œsophagus aid in propelling the food downward toward the stomach?

What is the first step in the process of digestion?

What terms denote digestion in the mouth?

What is the next stage of the process of digestion after deglutition?

What is meant by bolting the food? What effect has it upon digestion?

Do all the salivary glands pour out the same kind of fluid?

What chemical change takes place in food in the mouth?

What is the ferment called which produces this change?

What other fluid in the intestine has a similar effect on food?

Where does action take place on this fluid?

What other ferments do we find in the alimentary canal?

How is deglutition controlled by the nervous system?

At what stage does the involuntary nervous system control it?

How long does food remain in the stomach?

What relation do the stomach and intestines of man bear to those of the carnivora and herbivora?

Which of these two classes requires the shortest and simplest canal for digestive purposes? Why?

Why is the apparatus long and complicated in the herbivorous animal?

What class of animals has a number of stomachs? How many?

What is the shape of the human stomach?

What relation does it bear to the chest?

What muscle separates the two?

How many openings has the stomach?

At which side does the food pass from the stomach into the intestine?

Which is the cardiac extremity? The splenic extremity? The pyloric extremity?

What kind of membrane lines the stomach?

What kind of fluid does it secrete?

What is the appearance of this membrane?

How far does it extend in the alimentary canal?

What effect has the mucus upon this membrane?

What fluid is secreted in the stomach when it is empty?

What fluid is poured into the stomach when food is introduced into it?

Of what ingredients is the gastric juice composed?

What is the acid in the gastric juice?

By what means are the gastric juices secreted?

What is the shape of the glands secreting the gastric juice proper?

At which end of the stomach is it most largely secreted?

What is the appearance of the interior surface of the stomach?

What action is exerted by the muscular coat of the stomach?

What is the arrangement of its muscular fibres?

When all these fibres contract together, what effect have they upon the stomach?

When the longitudinal fibres act alone, what is the result?

When the circular fibres act alone, what effect have they on the stomach?

How is the shape of the stomach affected by the contraction of the oblique fibres?

What kind of membrane is the outer coat of the stomach?

In the herbivorous animal with four stomachs, which one of these resembles the human stomach?

What is rumination?

Which stomachs of animals are interested in it?

What is the arrangement for digestion in birds?

What is the gizzard, and in what class of birds is it chiefly developed? What is its action?

What is the digestive arrangement in insects?

In carnivorous birds what is the form of the digestive apparatus? What action takes place in the stomach?

What is the peculiarity of the digestive organs in reptiles? In fishes?

What is the nature of the food of insects?

What is the arrangement of the digestive organs in herbivorous insects? In carnivorous?

What is the simplest form of stomach in the lower animals?

What change takes place in the stomach when food enters it?

In what direction does the dilatation take place?

What kind of movement does the stomach undergo?

How may respiration be affected by it?

What is the result of the regular contraction and dilatation of the stomach in digestion?

Describe the churning movement of that organ.

How long a time is occupied by each revolution?

What effect has the diaphragm in digestion?

What is meant by the vermicular movement of the stomach and intestines?

What are the two important features of stomachal digestion?

What is chyme?

Is it always the same fluid?

How is its composition affected?

Is any action exerted by the saliva on the contents of the stomach?

What classes of substances are digested in the stomach?

How may acids be developed in the stomach?

Where do oily matters chiefly undergo digestion? Vegetable food?

Where are thick fluids, as soup, acted upon?

Why is animal food more digestible than vegetable food?

What effect have animal fats and oils on digestion?

Where are bread, potatoes, and pastry digested?

What becomes of thin liquids taken into the stomach?

How long a time does digestion in the stomach occupy?

Under what circumstances is this period prolonged?

To what temperature is food exposed in the stomach?

What change is effected in albuminous matters under the influence of the gastric juice, acids, and heat?

What is the active medium through which it is effected?

What is meant by the antiseptic action of the gastric juice?

What may be summed up as the result of stomachal digestion?

Through what vein do they pass to the liver?

Where does this blood go after leaving the liver?

Why is the effect of alcoholic drinks so rapid?

Why is the liver affected in drunkards?

When are substances said to be digestible?

What relation does digestibility bear to the nutritive qualities of articles of food?

What is learned from tables of digestibility of food?

What is said of the digestibility of milk?

What quantity of gastric juice is secreted in the twenty-four hours?

How is healthy digestion affected by mental or physical exercise?

What precautions should be taken after meals?

How is digestion impaired in early life?

What is the condition called "bilious," and how is it excited?

What rule should be adopted in regard to taking fluids at meals?

What error of diet is committed in this respect?

What is the effect of water on the gastric juice?

What facts do we learn from the table of digestibility, as it is called?

Into what part of the alimentary canal does the stomach empty?

Into what two great divisions are the intestines divided?

What is the length of the small intestine?

Into what parts is it divided?

Why is the duodenum so called?   The jejunum?   The ileum?

Is intestinal digestion effected in the small or large intestine?

What is the use of the large intestine?

Into what parts is the large intestine divided?

Why is the cæcum so called?

What are the coats covering the intestines?

What is the serous coat called?

What kind of movement takes place in the intestines?

Is it voluntary or involuntary?

What effect is produced by contraction of the circular fibres?

What is the effect of contraction of the longitudinal fibres?

What is the appendix of the cæcum, and what is its use?

What peculiarity does the cæcum present in grain-eating birds?

What peculiar arrangement of the intestines in reptiles? In insects?

What is the appearance of the lining membrane of the intestines in man?

What is the arrangement for absorption?

What fluid is poured out on the mucous membrane?

How is it secreted?

After what anatomists are the intestinal glands named?

What two important fluids are poured into the duodenum?

What is the total amount of digestive fluids secreted in the twenty-four hours?

How much saliva? Bile? Gastric juice? Pancreatic juice? Intestinal juice?

Which of these fluids is alkaline?

What is the action of the pylorus?

Is it under the control of the will?

With what juices is the food mixed in the duodenum?

What effect has the pancreatic fluid on oily or fatty matters?

What effect on starchy matters?

What organ secretes the bile?

What is the situation of the liver, its weight and width?

Is all the bile used for digestive purposes?

On what portions of the food does the bile act?

What organ contains the bile, and what connection has it with the liver?

What effect has digestion in the stomach upon the bile?

What effect has bile on the coats of the intestine?

What is the peristaltic action of the intestines?

What is chyle?

What is the process called by which it is formed?

What are the villi?

What vessels take their origin from them?

Why are they called lacteals?

What effect has the intestinal juice on the food?

Is the intestinal juice acid or alkaline?

How does it prevent fermentation of the food?

When may digestion proper be said to be terminated?

To what is the presence of air in the intestine usually due?

To sum up the steps in the process of digestion, where does action take place on starchy matters? On fatty matters? On albuminous matters?

ABSORPTION.

Various Forms of Absorption.—Having completed the study of the digestion of food, we now consider the process by which it is converted to the uses of the system. Under the subject of Absorption, in other words, we describe the process by which alimentary and other substances are introduced into the blood. The absorption of thin fluids by the bloodvessels in the stomach, already referred to, is an illustration; and the veins of the intestine have similar powers of absorption. Allusion has also been made to the arrangement of villi in the small intestine, containing lacteal (so called from their milky appearance) or chyliferous or chyle-bearing vessels and bloodvessels, both of which absorb fluids. Another form of absorbent vessel is that which is engaged everywhere in removing worn-out parts which have done their duty, so to speak. These are very numerous, and are called the *lymphat'ic vessels*, the fluid

Villi.

Tubular glands.

Areolar tissue.

Circular muscular fibres.

Longitudinal muscular fibres.

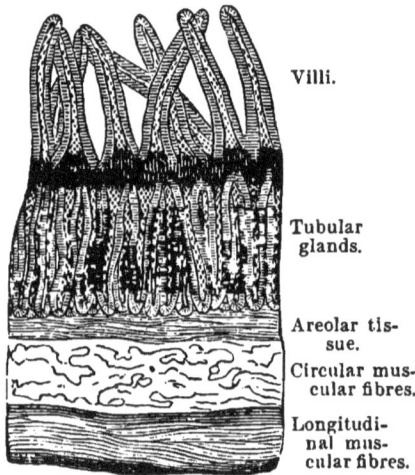

FIG. 42.—SECTION OF THE SMALL INTESTINE, AS SEEN UNDER THE MICROSCOPE.

120

they contain being called *lymph*, and this is also carried directly into the blood.

Intestinal Absorption.—The lining membrane of the small intestine, which we have already stated to be a mucous membrane, is thrown into folds, so that the extent of its surface is very much increased, in reality doubled. A much larger space is thus obtained for absorption than would otherwise exist. In addition to this increased surface there are *villi*, or delicate elevations on the surface of the mucous membrane of the intestine, so closely placed together as to give it a fine velvety appearance (Fig. 42). In each villus, as already stated, is a network of small vessels, thoroughly supplied with blood, and a chyliferous vessel (Fig. 43). The nutritious portions of the food are absorbed into the villi through a process which is generally known as *imbibition*, which has already been referred to under the subject of Digestion. These villi dip into the cavity of the intestine, and come in contact with the materials that have already been digested.

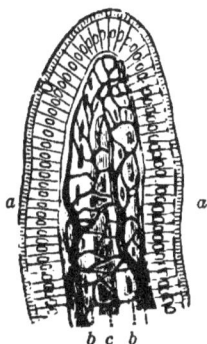

FIG. 43.—ONE OF THE VILLI OF THE INTESTINE, AS SEEN UNDER THE MICROSCOPE.

a, layer of cells; *b*, artery and vein; *c*, commencement of lacteal vessel.

It is well known that when two different liquids—milk and water, for example—are separated from one another by a membrane, two currents are established in opposite directions through the pores of this membrane, but in different degrees. The experiment can be shown by placing milk in a tightly-closed bladder and suspending it in a goblet of water; it will be found that the water will gradually become milky, and the milk will become thoroughly diluted with the water. The same condition exists

11

in the intestines.   The lacteal vessels do not have open
mouths to suck up the chyle, but absorb it through the
membrane covering them, on the principle just mentioned.
After entering the lacteals, which are really a part of

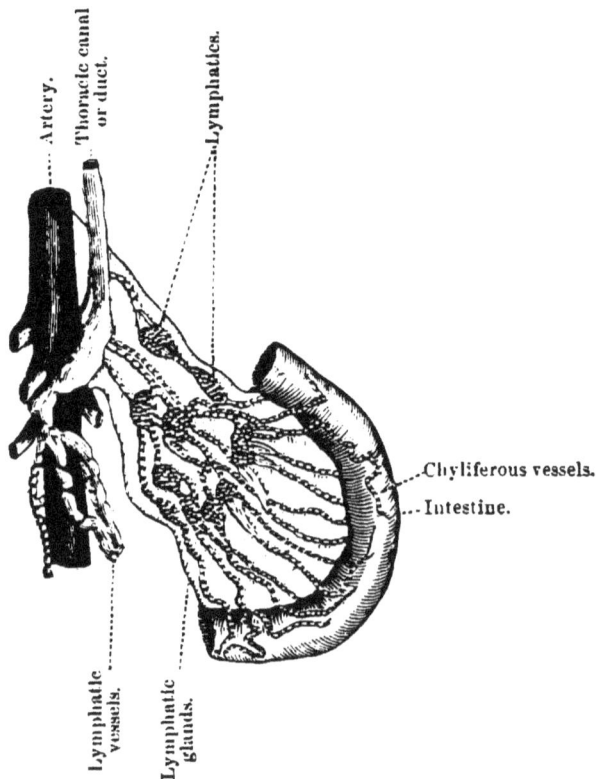

FIG. 44.—LACTEAL OR CHYLIFEROUS VESSELS OF THE INTESTINE.

the lymphatic system, and contain lymph when digestion
is not going on, the chyle is emptied into the *thoracic canal*
or *duct* (Figs. 44, 45, 46), a long canal which passes along
the back part of the chest, in front of the spinal column,
and carries the chyle up to the left side of the neck to
empty it into one of the large bloodvessels in that region.
On the way between the intestine and the thoracic canal it

passes through a number of small bodies, called *lymphatic glands* (Figs. 44, 45, 46), which produce changes upon it that render it more like the blood with which it is to be mixed. When once mixed with the blood, it passes with that fluid to the heart.

To sum up, therefore, the action of absorption of chyle, we may repeat that it is absorbed either, if very thin, through the bloodvessels of the intestine, passing to the

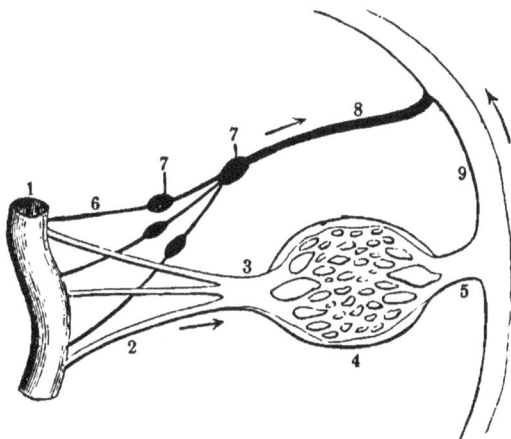

FIG. 45.—GENERAL PLAN OF DIGESTIVE ABSORPTION.

1, intestine; 2, bloodvessels: 3, portal vein; 4, liver; 5, veins of liver; 6, chyliferous vessels; 7, lymphatic glands; 8, thoracic duct; 9, venous system.

liver through the portal vein, or by the lacteals and the thoracic duct to the veins of the neck. (This is clearly exhibited in Fig. 45, which is intended to be an imaginary outline plan or model of the parts represented in Fig. 44.) The chyle becomes more and more unlike itself and more like the circulating blood as it passes along with that fluid for the nourishment of the body. It will be remembered that various changes were effected on food in the process of digestion in the stomach and intestines. The work of the chyliferous vessels is to absorb the fatty matters which we have shown to be made into an emulsion by the juices of the intestine; those albuminous matters which resulted from the action of the gastric juice on meats; and such portions of sugar as have been converted from starchy matters. The veins of the intestine also absorb some of

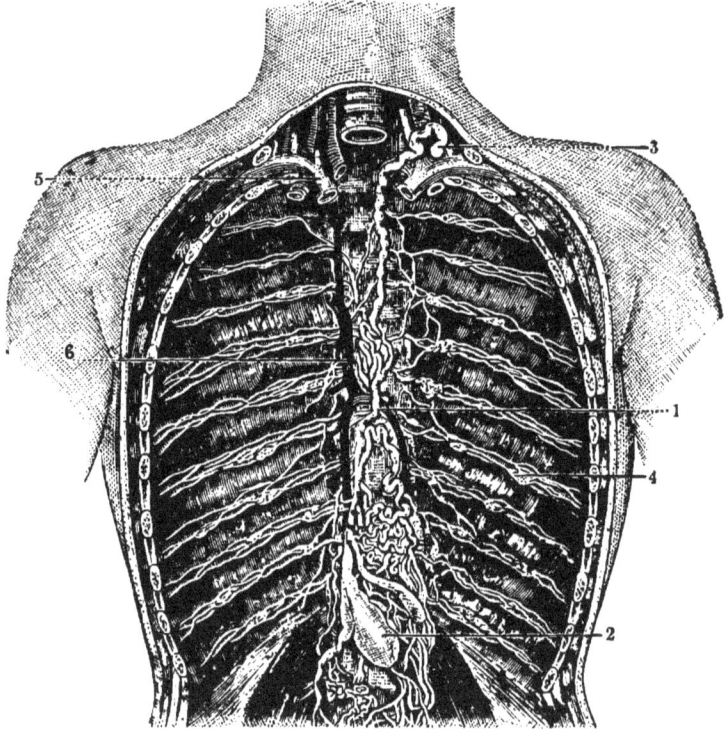

FIG. 46.—THORACIC DUCT AND CHYLIFEROUS VESSELS.

1, thoracic duct; 2, its lower part; 3, termination in the vein; 4, lymphatic ganglions; 5, 6, veins of the right side.

the soluble products of digestion, but not the fatty portions. It must not be forgotten, too, that the veins of the stomach also absorb thin fluids in a similar way.

Lymphatics.—This system of vessels penetrates almost all parts of the body, receiving by absorption the various materials collected from the wear and tear of the system.

The vessels are on the surface or deep-seated (Figs. 47, 48). After these materials are elaborated in the little bodies called lymphatic glands, which are found everywhere in the course of these vessels, the fluid resulting, which is colorless and transparent, is carried to be emptied into the blood. These lymphatic or absorbent vessels are very numerous in almost every part of the body, and run into one another like streams into a river, gradually forming vessels of considerable size, which finally unite and either empty into the thoracic duct, in front of and parallel to the spinal column, or on the right side of the body into a large canal called the right lymphatic trunk. Both of these empty the lymph into the current of blood. At various points in the interior of the lymphatic vessels there are projections called *valves*, which prevent the lymph from flowing backward in its course. The arrangement is similar to that of the veins, which we shall hereafter describe and illustrate.

The whole plan of lymphatic vessels and lymphatic glands is a system of drainage similar to that employed in some soils, in which

FIG. 47.—LYMPHATIC VESSELS ON THE SURFACE OF THE ARM.

the water drained from boggy lands forms channels, which unite to make larger streams, still increasing in size, and

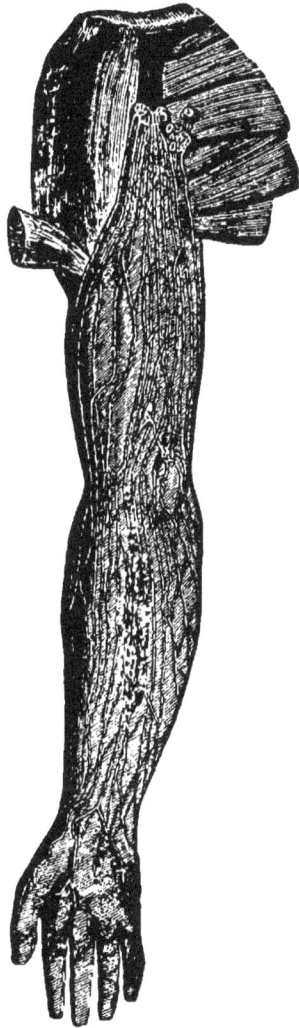

11 *

occasionally expanding into pools, where fresh supplies of material are absorbed, and at last pouring into the river or open sea. The fluid is carried onward in the lymphatic vessels by the powers of contraction of these vessels and by the presence of valves, which prevent the lymph from flowing backward. When the muscles engaged in breathing, walking, etc. contract or expand, they press upon these vessels in all parts of the body, and assist the flow of fluid in them. The breathing of air outward and inward during respiration also exerts a similar influence.

In some animals, as the frog, there is an enlargement, with muscular fibres connected with it, called a *lymphatic heart*, which propels the lymph very much in the same way that the heart itself maintains the circulation of the blood. In fishes and reptiles the lymphatic vessels are of larger size, proportionately, than they are in man or birds. Birds have two thoracic ducts. The lymphatic glands are more numerous in man and the upper classes of animals than in any others. In those animals which are called inver'tebrate because they have no vertebra or backbone, as the shell-fish, spiders, etc., there are no lymphatic or chyliferous vessels. In some of these the digested food at once becomes blood;

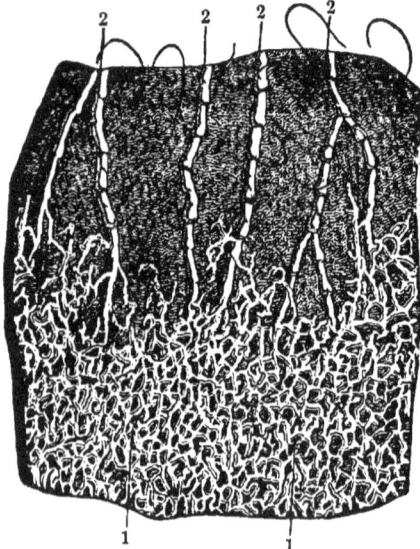

FIG. 48.—DEEP LYMPHATICS OF THE FINGER.

1, 1, deep network of lymphatic vessels of skin; 2, 2, lymphatic trunks connected with these vessels.

in others, which have an imperfect apparatus for circulation, it passes through the coats of the intestines, and is carried into the bloodvessels by endosmose. In insects the fluid is diffused without passing into circulating vessels.

QUESTIONS.

What is meant by absorption?

What relation does it bear to digestion?

What fluids are absorbed by the blood-vessels of the stomach and intestines?

What other vessels are found in the intestines?

What are the lymphatic vessels, and what do they contain?

How is the absorbing surface of the intestine increased?

What are the villi?

How is absorption through the villi effected?

Explain the process of endosmose or imbibition.

What becomes of the chyle after entering the lacteals?

What little bodies found along these vessels assist in the elaboration of that fluid?

Through what channel does chyle pass to the liver?

What kinds of materials are absorbed by the chyliferous vessels after digestion?

What action is exerted by the veins of the intestine?

Where are the lymphatic vessels found?

What is their duty, and where is the lymph emptied?

What arrangement prevents the backward flow of the lymph?

What other forces assist in the movement of the lymph in the vessels?

What is the lymphatic heart?

What is the peculiarity of the lymphatic system in fishes and reptiles? In birds?

What kinds of animals have no lymphatic system? What then becomes of the digested food?

What is the arrangement in insects?

RESPIRATION.

Object of Respiration.—It has now been shown in what way the fluids formed during digestion, and the lymph resulting from the general wear and tear of the various parts of the body, pass into the general circulation. When, however, they empty into the veins they are not in a fit condition to nourish the tissues. The blood itself is not then perfect; it must be purified and its tone improved. To effect such a change upon it it is necessary that it should be brought in contact with the air breathed into the lungs, and this action begins at the very first moment of existence. The mode in which this conversion of venous blood into arterial blood is produced is now to be studied under the head of Respiration.

Changes Produced.—Respiration may be defined as the function by which venous blood is converted into arterial blood. The change takes place in the lungs, the oxygen of the air being imparted to the blood and carbonic acid being given off.

To show clearly to the eye that carbonic acid is given off from the lungs, the following simple experiment may be made: Fill a wineglass with lime-water, and breathe into it through a glass tube, when the invisible carbonic acid of the breath will at once unite with the lime and soon throw down flakes of chalk or carbonate of lime. After obtaining this deposit, add a little vinegar to it—which has a stronger attachment to lime than carbonic acid has—and

123

the carbonic acid will be set free, and be seen effervescing, as it is called, or bubbling up. There is also a certain amount of watery vapor discharged from the breath. It is seen as condensed on a window-pane when breathed upon, or on a cold day in the cloudy vapor that is visible as it comes from the mouth and nose.

The process of respiration occurs by a passage of gases —oxygen and carbonic acid—through the delicate membrane of the bloodvessels of the lungs, just as we have already seen was the case with fluids in the absorption of chyle. It has been found by experiment that even when an already moistened membrane will not admit of the passage of fluids through it, gases can frequently penetrate it. Before the blood reaches the lungs it is called *ve'nous blood*, because it flows through the veins, and is of a dark bluish-red color. After it has received oxygen in the lungs the color is changed to a bright vermilion, and the blood is then called *arte'rial blood*, because it is carried in the arteries.

One peculiarity in regard to the act of respiration which cannot be stated of the other functions we have considered —Digestion and Absorption—is that it cannot be wholly suspended or stopped for more than a few moments without causing immediate death. Venous blood requires the conversion just mentioned, because it is an impure fluid containing materials that have already served for the support of life in various parts of the body. Even the chyle, that is to play such an important part in maintaining the general nutrition of the body, must be made more perfect, more like the blood itself.

When we come to study the subject of the Circulation we shall be able to appreciate more clearly the manner in which the blood passes from the right side of the heart to

I

receive its supply of oxygen. For the present we need only impress the general fact that such a conversion does take place, and defer the consideration of the details as a part of the sketch of the general circulation of the blood. It may be stated that the subject of Respiration includes only this passage through the lungs, for when the blood leaves the lungs after being aërated, it is more appropriately studied under the head of the Circulation.

The Chest and its Contents.—Before studying the process of respiration, let us first inquire into the nature of the organs concerned in it. The *lungs* are two large marble-blue organs, which with the heart fill the whole cavity of the chest, and which when once filled with air float in water ever afterward, as may be shown with a calf's lung removed from the animal. The chest, technically called the *tho'rax*, is a conical bony cage (Fig. 1), covered on the outside and lined on the inside with muscles, and separated from the abdomen by a large muscle called the *diaphragm*, which forms the floor of the chest and separates it from the abdomen. The diaphragm is so called because it separates these two important cavities, in this respect acting like a partition or diaphragm anywhere else. The thorax is made up of a portion of the spinal column at the back part and twelve ribs on each side, passing from the spine to or toward the breast-bone (Fig. 1). Seven of the ribs are attached to it—not directly, but by the medium of elastic material called cartilage,—while the others are either fastened to one another or float unattached at their front part.

The ribs are made very movable, so that they can rise and fall, and in this way dilate and contract the chest during respiration. Muscles pass between the ribs, and by their action upon them assist in respiration. Perhaps the most important muscle concerned in this process is the diaphragm, which is attached around the base of the chest, and when relaxed forms an arch, the middle of which is opposite the lower end of the breast-bone. All the muscles of the chest and also those of the abdomen take part in respiration, and, when breathing is rapid and excited, those which raise and lower the ribs are especially called into play.

The Lungs.—The lungs are so arranged as to give a very

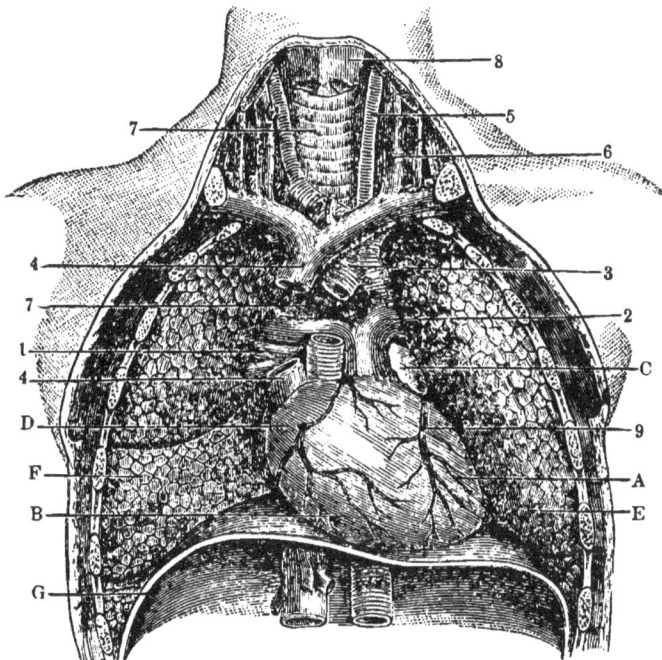

FIG. 49.—LUNGS, HEART, AND DIAPHRAGM IN POSITION.

1 pulmonary vein; 2, pulmonary artery; 3, main artery from heart; 4, vein; 5, carotid artery; 6, jugular vein ; 7, windpipe; 8, larnyx; 9, coronary artery; A, B, C, D, heart; E, F, lungs; G, diaphragm.

large surface for the contact of the blood and the air, and to do this to the best advantage each of the very small bloodvessels is completely surrounded by air. The lungs are divided into smaller portions, called lobes—the right lung into three, the left into two. A delicate membrane, called the *pleu'ra*, lines the interior of the chest and is reflected over the lungs and heart. It secretes a fluid which keeps it always moist. In infancy the lungs are pale red, but get darker by age, being in old persons a livid blue. The air of the external atmosphere is tempered on its way through the air-passages, especially by the warm circulating

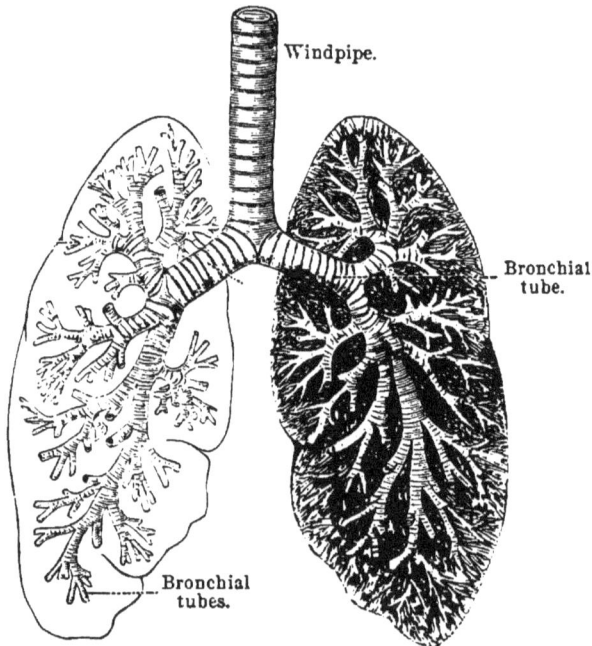

Windpipe.

Bronchial tube.

Bronchial tubes.

FIG. 50.—THE LUNGS AND BRONCHIAL TUBES (the lung-structure on one side being supposed to be removed).

blood in the lungs, so that when it comes in contact with the delicate structures of the latter it produces no injurious effect. The air after it enters the mouth passes along the throat into the *lar'ynx* (Fig. 83), which we shall hereafter describe as the organ of voice, thence into the windpipe or *trache'a* (Figs. 49, 50), which divides into two large tubes or canals called the *bronch'ial tubes*. These tubes divide, one

entering each lung, and subdivide until they become very minute and very nu-
merous, and penetrate every part of the structure of the lung (Fig. 50), leading to minute cavities, which terminate at last in blind extremities or sacs, having very thin walls, and called *air-cells* (Fig. 51). These cells are covered with a network of very minute bloodvessels. The air we breathe does not therefore circulate in the substance of the lungs proper, for it cannot get

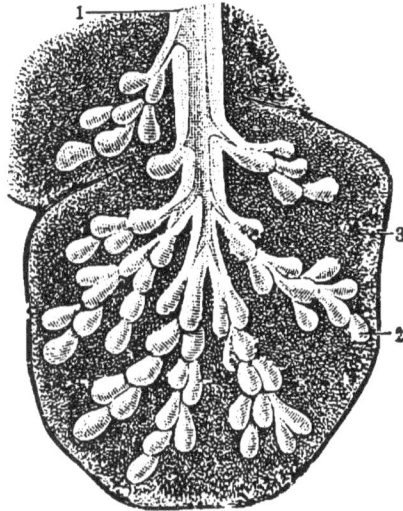

FIG. 51—LOBULE OF LUNG (magnified).
1, small bronchial tube ; 2, termination in air-cell; 3, pulmonary tissue.

beyond the air-cells just mentioned (Fig. 51). These become distended after a full inspiration of air, and remain partially so during life as long as breathing lasts. The air-passages in some portions have rings of cartilage (Fig. 50) to strengthen them and to impart elasticity. The air-passages are lined by a mucous membrane, similar to that which lines the alimentary canal, and to which reference was made in the chapter on Digestion ; and the same kind of fluid is poured out to keep the air-tubes moist. On this membrane, as seen under the microscope, are numerous cells like hairs, to which the name *cil'ia* (meaning eyelashes) has been given, and they have a waving motion like a field of grain, the motion being upward and outward toward the mouth. They possibly assist in carrying, in a direction away from the lungs, any fine particles of dust, etc. that might otherwise pass into the more delicate air-passages.

12

The Act of Breathing.—The terms *inspiration* and *expiration* are applied to the filling and emptying of the lungs in the act of breathing. These are effected by the alternate contraction and enlargement of the chest by the action of muscles, especially those connected with the ribs When the thorax or chest contracts, expiration occurs, and the air is expelled from the lungs through the various tubes already mentioned, and through the mouth and nostrils. Expansion of the chest—that is to say, its enlargement in length, width, and height—produces inspiration. The process of respiration is partially and but temporarily under the control of the will, so far as moving the chest is concerned, and it continues without cessation as long as life lasts. When expiration has taken place, the need of immediate inspiration becomes so absolutely felt that it cannot be postponed. Swimmers in making a plunge into the water cannot long delay their return to the air above to complete the act of respiration. There is, however, a slight but brief period of repose, which cannot be safely prolonged.

The alternate expansion and contraction of the thorax may be illustrated by a pair of bellows, the nozzle being the windpipe, the flexible leather uniting the boards the diaphragm, the boards themselves being the ribs, and the hinges the attachment of the ribs to the spinal column. When the diaphragm is extended, and the ribs elevated by the muscles connected with them, the dilated chest has its counterpart in the bellows when the boards are separated. The air rushes into the nozzle, as it would into the windpipe, to fill up the vacuum, and the bellows is emptied in a similar way to that of expiration. It may be said that the lungs are active in inspiration, but in expiration they are passive, their only duty being to empty themselves gradually by their own elasticity.

Gentle Respiration.—Gentle, easy respiration in man is effected by the elevation and depression of the diaphragm, without calling all the other muscles into active play. This muscle of course presses upon the contents of the abdomen whenever it is depressed in producing enlargement of the chest. When the diaphragm is elevated, as in expiration, the stomach and intestines again return to their natural position, and this explains the visible swelling and contraction of the abdomen during respiration. The effect of easy respiration by the diaphragm and abdominal muscles, as well as that of forced respiration by the chest, may be seen in the illustration (Fig. 52). Such easy respiration is that seen in those who are sleeping. In deeper and more active respiration, such as that occurring under excitement, a large number of other muscles are called into exercise.

FIG. 52.—GENTLE AND FORCED RESPIRATION (the latter shown in the dotted line).

Number of Respirations.—As a general rule, the number of respirations is eighteen in a minute; that is to say, there are accomplished in this period of time both inspiration and expiration, with a very brief period of repose. The estimate usually made is that while the heart is making four beats in the circulation of the blood one act of respiration is taking place. Of course during exercise and motion the number of respirations is increased. Age has also a modifying effect; in the infant the number is considerably greater than in the adult. A young baby at and soon after birth will breathe at least forty times a minute; the child of five years of age will have about twenty-five respirations a minute; and so the number goes on diminishing until about the fifteenth or sixteenth year, when the number reaches, say, eighteen per minute—characteristic of adult life. It may be briefly said that whatever excites the individual in any way, especially if it affects the action of the heart, increases the activity of respiration, and therefore the number of respirations.

Full Respiration.—A greater amount of muscular effort is used than in ordinary easy respiration when we breathe rapidly or forcibly. The muscles of the chest, even those which are attached to the neck or the arm, are called into service so as to enlarge that cavity. The upper part of the chest has usually but little motion, unless the breathing be rapid, especially in the female. Indeed, woman breathes more with the chest, even in gentle, easy respiration, than man. This natural tendency is increased by the mode of dress, which should, however, be such as not to compress the walls of the chest or to interfere with the free motion of the ribs. It has been shown by experiment that in easy inspiration the larynx, the organ of the voice, is not movable, and there being perfect absence of fatigue, singing can be prolonged for a greater length of time. When the upper ribs are rapidly raised, as in forced respiration, general and vocal fatigue ensue very rapidly, for the organ of voice can no longer depend on the assistance of surrounding muscles, but must call into play all its own muscular power.

Sounds of the Chest.—In applying the ear to the chest a gentle sound is heard, caused by the penetration of air into the air-cells. This sound, as well as that louder and rougher sound caused by the passage of air along the various tubes, is changed in its character by disease of those parts; and the physician, who is familiar with the natural, healthy sounds in breathing, soon recognizes the differences noticeable in them when so affected by inflammation or other diseased condition. He can also learn by striking over the chest with his finger whether the air-cells are doing their proper duty, for the sound will then be clear, as it will always be in sounding over a cavity filled with air. If the cells do not contain air, but have become solid or obstructed from any cause, the sound will not be any longer clear, but dull and obscured.

Capacity of the Lungs.—When air enters the air-cells of the lungs it is not wholly expelled in expiration. A certain quantity always remains behind, and this is called the *reserve air*. It is hardly, therefore, correct to speak of emptying the lungs in the act of expiration. When the lungs are fully expanded by the air, the quantity has been estimated at rather more than three hundred cubic inches, or more than four quarts. The average quantity breathed out at each expiration is estimated at about twenty cubic inches, so that rather less than three hundred cubic inches remain behind after each expiration. The entire air in the lungs is therefore thoroughly renewed about once in a minute. In easy respiration the whole amount of the breathing capacity of the lungs is not called into play. Estimating the quantity of air breathed per minute at three hundred and sixty cubic inches, or about ten pints, an idea may be formed of the immense quantity breathed in the course of twenty-four hours.

The Air-Cells.—To give some idea of the size of the air-cells, it may be stated that the smallest bronchial tubes—capillary tubes they are called, because they are almost hair-like in their dimensions—measure from the fiftieth to the thirtieth of an inch, the air-cells from the two-hundredth to the seventieth part of an inch. A thick network of delicate and minute bloodvessels covers the walls

12 *

of the air-cells and the passages between the cells. These communicate on one side with the arteries in the lungs, called the pulmonary arteries (from *pulmo*, "a lung"), on the other with the pulmonary veins ; and here the change from venous to arterial blood takes place, the pulmonary veins carrying the purified blood to the left side of the heart (Fig. 152).

The Air we Breathe.—Pure atmospheric air is composed of two gaseous substances, called nitrogen and oxygen, mixed together in the proportion of 79 parts of nitrogen to 21 parts of oxygen, or, in round terms, one part of oxygen to four of nitrogen. It contains in addition a small quantity of watery vapor and a very minute quantity of carbonic acid—only one part in 2000 parts of the atmosphere—scarcely sufficient to take note of. Oxygen alone would be too stimulating, and life would be maintained at such a high pressure, so far as respiration is concerned, that death would soon occur from over-stimulation. Diluted with the nitrogen, however, it is a powerful and energetic agent for respiratory purposes. The amount of carbonic acid in the air under ordinary circumstances is not sufficient to do any harm, but when it becomes increased in quantity, as we shall hereafter see, very serious effects ensue. When the air is examined after it has been breathed out of the lungs, it is found to contain a much smaller amount of oxygen than when it was inhaled, and a greatly increased quantity of carbonic acid. This shows that oxygen has been absorbed into the lungs and carbonic acid given off from them. A small amount of watery vapor is also exhaled from the lungs at each inspiration.

It will be seen hereafter that the blood contains numerous little structures, called *corp'uscles* (signifying " little bodies," from *corpus*, " a body "), visible only under the microscope.

These are the parts of the blood that are acted upon by the oxygen taken into the lungs at an inspiration.

Mechanical Actions in Breathing.—A variety of mechanical actions are connected with the function of respiration. The exercise of the voice is an illustration, whether used in speaking or singing. Coughing, sneezing, laughing, smelling, sobbing, spitting, yawning, snoring, etc. are all modified forms of inspiration or expiration, or of both combined.

Sighing is a deep inspiration, by which air is slowly inhaled in large quantities, frequently because, through languor or emotion, the blood does not receive a sufficient amount of pure air, or, in other words, is not properly aërated in the lungs. The process of respiration may become sluggish and irregular if the mind is wholly absorbed by any cause diverting the attention from it, such as attends the preoccupation of the mind, as by literary labor, emotion in some form, etc. The blood does not then become perfectly aërated, and the poor venous blood does not undergo properly the change into the richer arterial blood. It is then that a long sigh or series of sighs comes to our aid, and rapidly supplies the amount of air necessary to give to the blood the oxygen it requires.

Coughing is a violent action of the muscles of expiration and contraction of the muscular fibres of the bronchial tubes. *Laughter* is a convulsive action of the muscles of respiration —including the diaphragm—and of the voice, combined with action of the muscles of expression of the face. *Yawning* is a deep inspiration, attended with contraction of the muscles of the lower jaw and part of the throat. *Sobbing* is similar in its causes to laughter. *Panting* is a series of short, quick inspirations and expirations, and seems to have for its object the rapid renewal of air in the lungs in cases in which the circulation is too rapid or where an extra supply of fresh air is demanded. *Smelling* is a series of short inspirations while the mouth is shut, so that the whole impression of the odor may be made upon the cavities of the nose alone. *Sneezing* is a violent expiration, in which the air, rapidly driven from the chest, agitates the cavities of the nose with a peculiar but familiar sound.

Obstacles to Respiration.—We have already said that the danger to life is great and immediate if the respiration be suspended for a few moments. This may occur from various causes, such as from strangulation, from drowning, from hanging, from pressure on the chest interfering with its movements, etc.; but whatever the exciting cause may be, the effect is always the same—the non-conversion of venous into arterial blood, the necessary oxygen not being received. This condition of the lungs may result from absence or insufficiency of the pure air that the lungs are in the habit of inhaling; from the presence in the air of some other gas instead of oxygen, etc. Air once breathed is unfitted for further respiration, being no longer pure. The amount of oxygen in it is found to be greatly lessened.

If too many persons are crowded together in a room the atmosphere becomes more and more vitiated by the breathing out of so much carbonic acid from so many lungs, and the effect on each individual soon becomes apparent, for carbonic acid is what is called an irrespirable or unbreathable gas. To ensure perfect respiration the air that is breathed must be renewed, and not allowed to be contaminated. One of the most familiar and frequently quoted illustrations of the effect of confinement in a close atmosphere is that of the "Black Hole of Calcutta," as it is generally known. In a war between the English and the people of a portion of India in the eighteenth century, 146 of the former were captured and imprisoned in a small room only twenty feet square, into which but little light and air penetrated, and the heat was intense. The next morning, after a night of dreadful suffering and an interval of only eight hours, all except twenty-three of the prisoners so confined were found to have died.

Death is not due alone to the additional amount of impurity given to the atmosphere under the circumstances just named; there is also a certain amount of animal or organic matter exhaled from the lungs of man and all other animals which rapidly putrefies and poisons the atmosphere. Amid all the comforts of modern homes danger to health or life may ensue from the very conveniences which surround their inmates; a burning gaslight in a closed room at night will evolve far more carbonic acid than would be given off from the lungs, and burning fires in stoves and heaters may be perpetually poisoning the air.

Ventilation.—It will be at once seen, after these allusions to the ill effects of impurities in the atmosphere, how necessary it is to adopt all proper means of prevention, so that the air of rooms occupied at the same time by a number of persons can be properly changed and kept pure. Schools, assembly-rooms, hospitals, etc. should have doors and windows and chimneys so adjusted that such a result can be perfectly accomplished, and that all mental effort, in schools especially, should not be hampered by deficiency of attention to the true principles of the physiology of respiration. It is not necessary to go into the details of the best form of construction of buildings to ensure perfect ventilation. Let it be remembered that the first and most important object is to give an outlet to the impure air, which as it becomes warm will rise, and an inlet to the pure external air; that cold air is not any purer than warm; and that it is not necessary to create a draught, a gentle current

being all that is required. Perhaps one of the best modes
of ventilation is the open fireplace, which creates a draught
up the chimney and carries off impurities through that
channel.

Summary.—We may briefly sum up our knowledge of
the process of respiration in man as follows: The air when
breathed loses some of its oxygen, but acquires carbonic
acid. The blood undergoes a change in color, which is
undoubtedly caused by contact with oxygen. The blood in
the lungs loses some of its carbonic acid. The oxygen of
the inspired air passes directly through the coats of the
vessels of the lungs. Aqueous vapor is also discharged
from the lungs.

Respiration in Animals generally.—The process of respiration in other animals than man has its peculiarities, the
lungs of birds, fishes, and insects differing greatly. The
process is similar to that of man, so far as the interchange
of gases is concerned between the blood and the atmosphere,
oxygen being taken in and carbonic acid given off. In the
lowest forms of animal life this takes place without the
presence of lungs, the whole external surface acting as a
medium for this purpose. In such a case the atmospheric
air penetrates their tissues and exerts its action upon the
fluids contained in them. In animals which have a thick
or hairy skin or hide this would be impossible, and their
lungs are usually sufficiently developed to carry on the process of respiration.

In *birds* the lungs are quite small and attached to the
chest. The pleura covers them on the under surface only.
A considerable part of the chest and also of the abdomen
is occupied by membranous air-cells, with large openings
communicating with the lungs (Fig. 53). In some birds the
bones also form receptacles for the air, the object being to

make the body so light that flight will be easy. Those which fly most rapidly and to the greatest heights—the eagle, for example—have these bony cells in the largest number. The air thus supplied enables the bird to fly and sing without the necessity of taking breath constantly. The

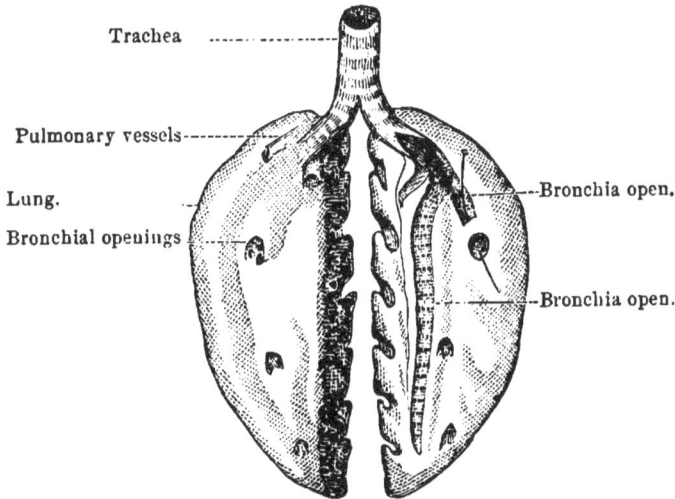

Trachea

Pulmonary vessels

Lung.

Bronchial openings

Bronchia open.

Bronchia open.

FIG. 53.—LUNGS OF A BIRD.

air seems to penetrate usually into bones which are intended for locomotion, as in the ostrich.

Respiration in *fishes* is performed by gills (technically called *branch'iæ*), membranes largely supplied with blood, placed behind the head on each side, to which is attached a movable gill-cover. Generally there are four gills on each side. The fish does not breathe the air of the atmosphere, but the air contained in the water in which it swims. The water passes into the throat, and is conveyed to the gills, through which it passes, making its way through the openings. Water is richer than the atmosphere in oxygen. Some fish, however, do not derive a sufficient amount of oxygen from the water, and rise occasionally to the surface for a larger supply.

Reptiles have feeble respiratory action; some have lungs; some early in life have gills; others are supplied with both lungs and gills. These last can live both on land and in water, and in some of them the whole surface of the body is an active medium for respiration and the interchange of gases. The lungs of reptiles receive air that is swallowed rather than breathed, and are generally made up of air-sacs divided by partitions.

In *insects*, respiration is usually effected through the

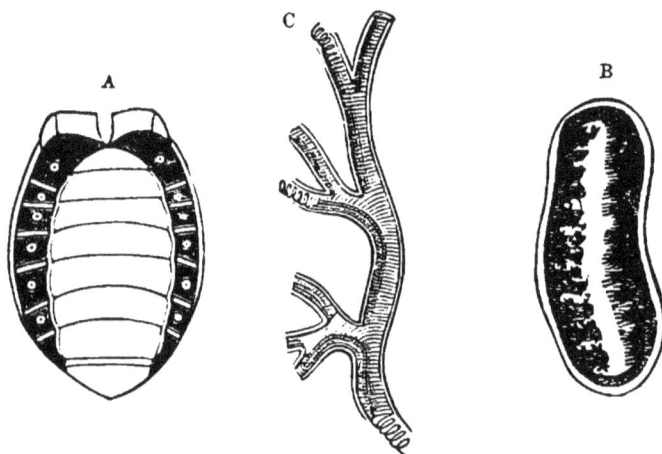

FIG. 54.—RESPIRATION IN INSECTS ILLUSTRATED.

A, stigmata, or respiratory orifices, of water-beetle; B, a single opening, greatly enlarged; C, trachea.

exterior of the body. In them the air enters through openings (technically called *stig'mata*) which are the terminations of air-tubes, or trach'eæ, and convey the air to all parts of the system. In all cases, however, there is the same kind of interchange of gases in animals generally that has been described as the chief feature of respiration in man. (See Fig. 54).

Respiration in the Vegetable.—There is a genuine process of respiration in the vegetable as in the animal, but of a different and peculiar kind. It is indeed a reversal of all that we have said in regard to the animal, for while the latter absorbs oxygen and gives off carbonic acid, vegetables, under the influence of. light, absorb carbonic acid and give off oxygen, the carbon becoming a part of the substance of the plant. Under other conditions oxygen is absorbed and carbonic acid gotten rid of by them, so that at night plants are unsafe companions in a sleeping-room. All those portions of the plant which are not green—and this applies to the flowers especially—absorb oxygen and give off carbonic acid, whether we examine them in the light of the sun or in the shade. Ripe fruit, it is said, respires in the same way, and grain that is germinating under the influence of air and moisture also has a similar interchange of gases. It may be stated, then, in brief, that the main difference between animals and vegetables, so far as respiration is concerned, is the fact that animals always absorb or take in oxygen, while vegetables do so only under certain conditions of their life.

QUESTIONS.

What change is necessary to purify the venous blood?

In what organ does the change take place?

What is the function called under which this conversion takes place?

What is the definition of Respiration?

What gases are concerned in the process?

Which is absorbed? Which is given off?

By what simple experiment can it be proven to the eye that carbonic acid is given off?

What gas is given off when a liquid effervesces?

What vapor is given off from the lungs, and when is it visible?

How is the passage of the two gases effected in the lungs?

Do gases or liquids penetrate moist membranes most readily?

What changes take place in the color of the blood during respiration?

Can respiration, like digestion, be suspended?

Why does venous blood require conversion?

What part of the circulation is concerned in the function of respiration?

What organs are the chief agents in respiration?

What organs fill up the cavity of the chest?

What is the thorax? How is it covered?

What muscle separates the chest from the abdomen?

How are the ribs arranged? How are they moved?

What other muscles take part in respiration?

What is the relation of bloodvessels to the air-tubes?

What is the pleura?

How does the color of the lungs vary at different periods of life?

How is the cold air tempered in breathing?

What is the course of the air when breathed?

What is the arrangement of the larger and smaller bronchial tubes?

What are the air-cells?

What two processes does the act of breathing include?

How is each affected by the motion of the ribs and the muscles?

What effect has expansion or contraction of the chest on inspiration or expiration?

How can the act of respiration be illustrated by the action of the bellows?

Are the lungs active in inspiration or expiration?

How is gentle, easy respiration effected?

What is the action of the diaphragm?

What is the number of respirations per minute?

How does this compare with the number of beats of the heart?

How does age affect the number of respirations?

What effect does excitement have upon them?

What other muscles are called into use in violent respiration?

How does the breathing of woman differ from that of man?

How is the voice affected by easy or rapid breathing?

What are the sounds of the chest in respiration?

What information do they afford the physician?

Are the lungs emptied when air is taken into them?

What is the breathing capacity of the lungs?

What is the average quantity breathed out at each expiration?

How often is the air in the lungs renewed?

What is the size of the smallest bronchial tubes?

What is the size of the air-cells?

What are the smallest bronchial tubes called?

What is the relation of the bloodvessels to the air-cells.

What are the main bloodvessels of the lungs called?

Of what materials is the atmosphere composed? In what proportions?

What vapor does the air contain?

How much carbonic acid is present in the atmosphere?

Why would not an atmosphere of oxygen alone do for respiratory purposes?

What little bodies exist in the blood?

What parts of the circulating blood are acted upon by the oxygen?

Mention some of the mechanical actions which are connected with the process of respiration.

What is sighing? In what conditions may it be beneficial?

Explain the respiratory actions involved in coughing; in laughing; in yawning; in sobbing; in panting; in sneezing; in smelling.

From what causes may respiration be suspended?

What is the effect of such suspension on the purification of the blood?

What is the effect of crowding persons together? With what gas is the air then contaminated?

What animal matter is given off from the lungs?

Sum up the general facts of the process of respiration.

How does respiration in animals resemble that of man?

Are lungs always present?

What is the arrangement of the organs of respiration in birds? How do the bones aid in the process?

What is the peculiarity of the respiration of fishes?

Which contains the most oxygen—air or water?

How is respiration effected in reptiles? In insects?

What are the stigmata of insects?

How does respiration in the vegetable differ from that of the animal? In what points do they resemble one another?

Why should plants not be kept in bedrooms at night?

CIRCULATION.

Object of the Circulation.—As the blood is the great medium through which all parts of the body are nourished, it becomes necessary that an apparatus should exist by which it may be distributed everywhere. We have already seen how blood was formed from the food and from the lymph as a result of the processes of Digestion and Absorption, and how, by the process of Respiration, it is purified in the lungs so as to be fitted for the nourishment of the body. Under the head of the Circulation we are now to study the manner in which it is to be transmitted to the various organs. In studying the course of the circulation we shall see that the purified blood, after leaving the lungs, passes directly to the left side of the heart. The heart is the great central organ of its distribution, and sends the aërated blood out upon its travels, to receive it once more after it has gone its rounds and performed its duty. The process is called Circulation, because the passage of blood from the heart all through the body and back to the heart through the bloodvessels is like a movement in a circle. The discovery of the plan of the circulation of the blood is due to a physician named William Harvey of London, England, who early in the seventeenth century demonstrated its existence.

The Heart.—The heart (Fig. 56) is chiefly a muscular organ, with wonderful contractile powers. It is in shape somewhat like a cone, and lies in the middle and front part of the chest, a little to the left (Fig. 55), between the

148

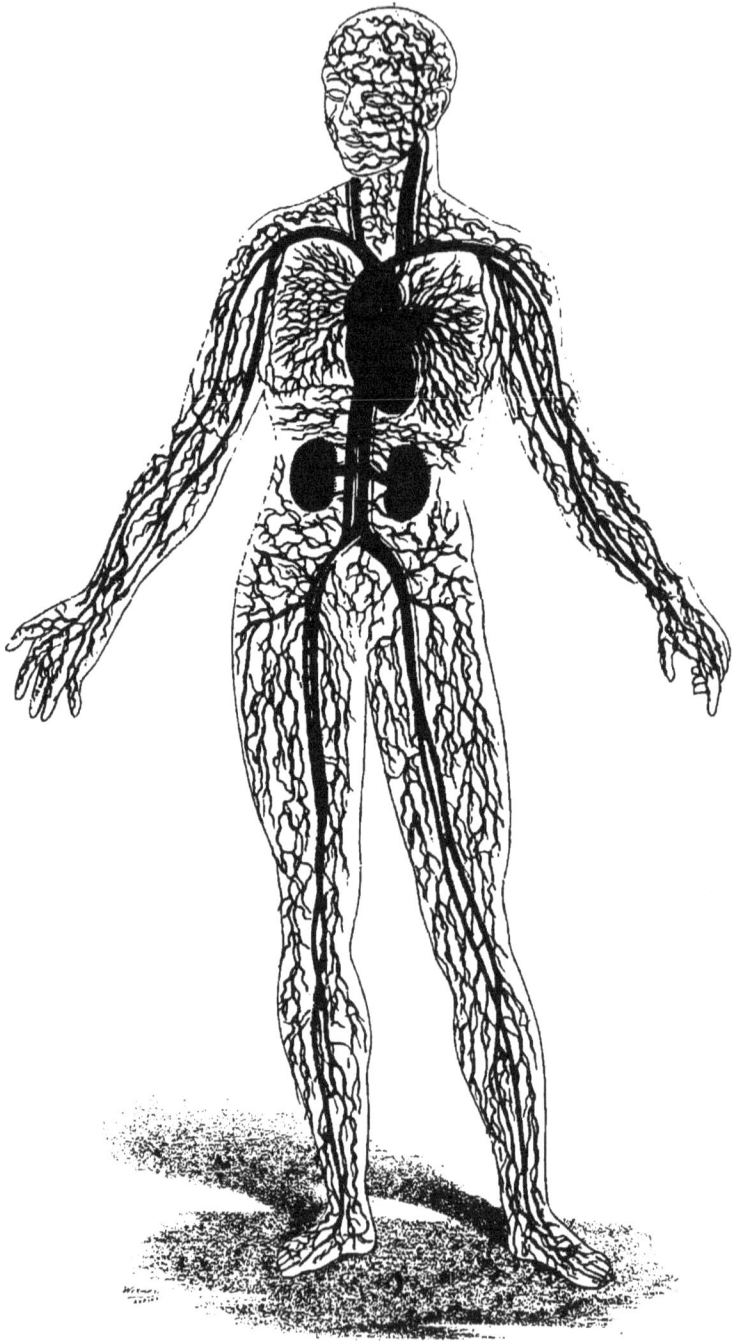

CIRCULATION OF THE BLOOD.

lungs (Fig. 49). It weighs usually about ten or twelve ounces, being, it is said, about the size of the fist, although this is a variable measurement. It is five inches long, three and a half wide, and two and a half thick. It is lined by a thin membrane, called the *endocar'dium* (from two Greek words meaning "within the heart"), and cov-

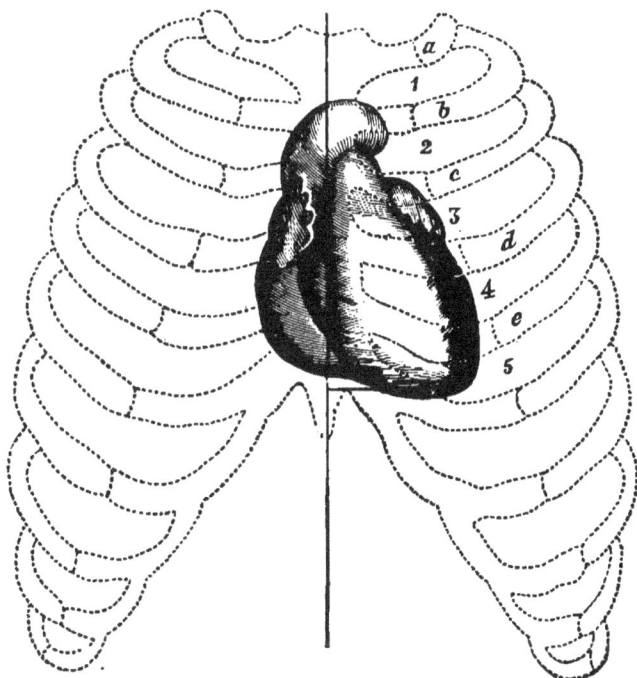

FIG. 55.—THE HEART IN ITS NATURAL POSITION IN THE CHEST.

a, b, c, d, e, ribs; 1, 2, 3, 4, 5, spaces between the ribs covered with muscles. (The vertical line represents the middle line of the body.)

ered by another called the *pericar'dium* (from two Greek words meaning "around the heart"). The heart is a red muscular mass, a familiar counterpart of which is seen in the bullock's heart on the butcher's stall.

The Heart a Double Organ.—The heart in man consists of four compartments or cavities, two of which receive the

13 *

blood and two propel it. The heart is really a double organ, having two distinct portions with their cavities. The right side, or right heart, as it might be called, receives the dark venous blood from the system generally,

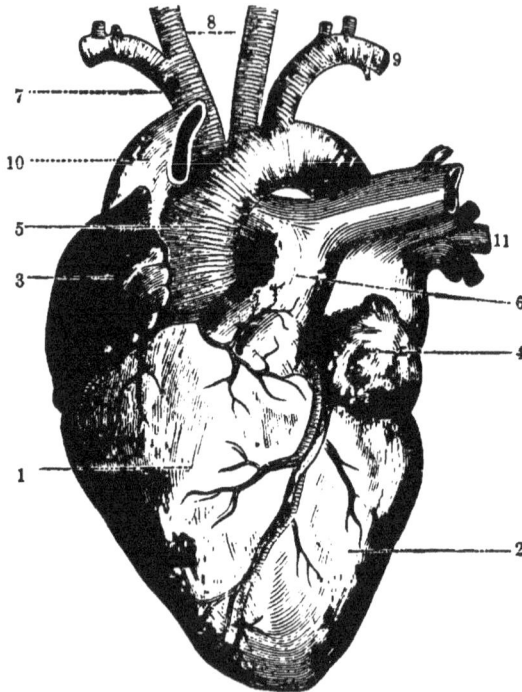

FIG. 56.—THE HEART, EXTERIOR VIEW.

1, right ventricle; 2, left ventricle; 3, right auricle; 4, left auricle; 5, aorta; 6, pulmonary artery; 7, 8, 9, large arteries branching off from aorta; 10, vena cava; 11, pulmonary veins.

and sends it to the lungs to be converted into red arterial blood. The left side, or left heart, receives the blood from the lungs, and sends it out everywhere through the body.

Cavities of the Heart.—Each side has two cavities, called an *aur'icle* ("a small ear") and a *ven'tricle* (literally, "a little stomach"), (Figs. 56–59). There is no direct communication between the right and the left heart except through the lungs. The right auricle receives the blood, and sends it into the right ventricle, which forwards it to the lungs. The left auricle receives the blood from the lungs, and the left ventricle propels it into large vessels, called *ar'teries*, to be distributed. As a rule almost without exception, the vessels that carry blood to the heart are called the *veins;* those which carry it in an opposite direction the arteries. The left side of the heart, having a greater amount of work to perform in the propelling of the blood through the whole system, has much thicker walls than the right side. The muscular walls of the ventricles on both sides are thicker than those of the auricles, as the duty of the former is to propel to a greater distance.

The Greater and Lesser Circulation. — The right heart, from its containing

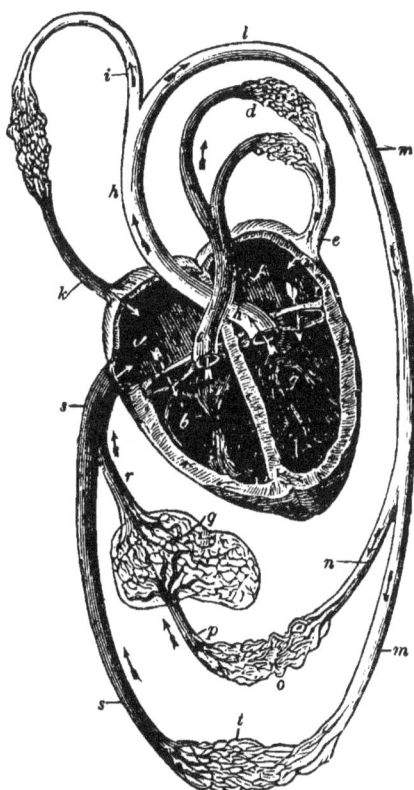

FIG. 57.—THE HEART AND ITS CAVITIES.
(Showing lesser and greater circulations.) a, right auricle; b, right ventricle, communicating through auriculo-ventricular opening; c, pulmonary artery, showing branches to each lung; d, capillary vessels of lesser or pulmonic circulation; e, pulmonary veins; f, left auricle, and g, left ventricle, communicating through left auriculo-ventricular opening; h, aorta; i, arteries; k, upper vena cava, bringing blood from upper portions of body to right auricle; l, arch of aorta; m, its descending portion; n, arteries of stomach and intestines; o, capillaries of intestines; p, portal canal; q, capillaries of portal system in liver; r, veins of liver; s, lower vena cava, bringing blood to right auricle from abdomen and lower portions of body; t, capillaries of greater or systemic circulation.

venous blood, is sometimes called the *ve'nous* heart, or the *pul'monary* heart because it carries the blood to the lungs. The left side is called also the *arte'rial heart,* because it contains arterial blood; the *aor'tic heart,* because it sends the blood directly into a large vessel called the aor'ta, the largest artery in the body; or the *system'ic heart,* because it distributes the blood to the system generally. This ar-

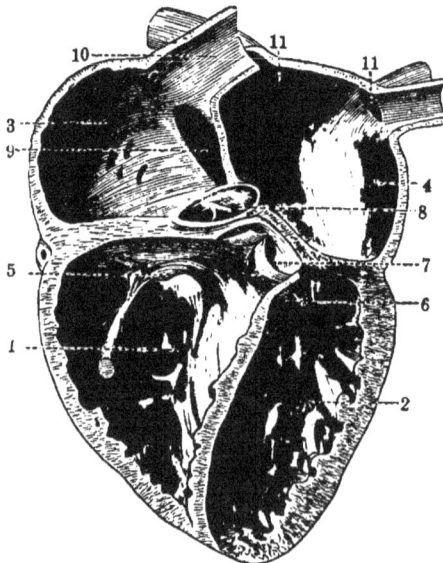

FIG. 58.—INTERIOR OF THE HEART.

1, right ventricle; 2, left ventricle; 3, right auricle; 4, left auricle; 5, opening between right auricle and ventricle—tricuspid valve; 6, opening between left auricle and ventricle; 7, pulmonary artery and semilunar valves; 8, origin of the aorta with its valves; 9, 10, opening of venæ cavæ into heart; 11, openings of pulmonary veins.

rangement of the heart establishes two circulations instead of a single one. One of these is from the right side of the heart, through the lungs to the left side, and is known as the *lesser or pulmon'ic circulation;* the other is that formed by the circuit of the blood all through the body from the left side of the heart, through the arteries and back by the veins, to the right side of the heart, and is called the *greater* or *system'ic circulation* (Fig. 57). The object of the smaller circulation, as already shown, is chiefly to reconstruct and purify the blood by aëration in the lungs; that of the greater circulation is the nourishment of the various organs.

Valves of the Heart (Fig. 58).—The opening between the right auricle and right ventricle is guarded by a valve called, from its shape, the *tricus'pid valve* (because it has three cusps or points), and that between the left auricle and left ventricle by the *mi'tral valve*, from its fancied resemblance to a bishop's mitre. These valves are intended to pre-vent the reflow of blood, so that the circulation shall be continuous. As the blood cannot flow backward when these valves are closed, it must go forward when the

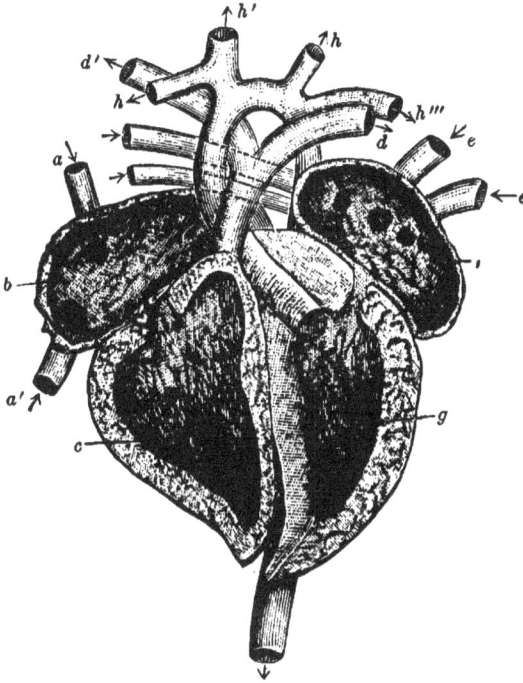

FIG. 59.—GENERAL VIEW OF THE HEART AND GREAT VESSELS PROCEEDING
FROM IT.

a, a', venæ cavæ : *b*, right auricle ; *c*, right ventricle ; *d, d'*, pulmonary arteries; *e, e'*, pulmonary veins ; *f*, left auricle ; *g*, left ventricle; *h, h', h'', h'''*, main arteries branching off from the aorta.

heart contracts upon its cavities. There are half-moon-shaped valves also, called therefore *semilu'nar*, both at the mouth of the pulmonary artery—a vessel which carries the blood from the right ventricle to the lungs—and at the

origin of the aorta, or main artery of the body, from the
left ventricle.

Movements of the Heart.—The movements of the heart
are rhythmical; that is, in a regular succession of contractions and relaxations. When the heart contracts the
act is called the *sys'tolē* (a Greek word meaning "contraction"), the relaxation or dilatation being known as the
dias'tolē (a Greek word meaning "dilatation"). The two
auricles contract and dilate simultaneously, and the two
ventricles undergo similar movements together; that is to
say, there is contraction and dilatation of the auricles, suc-
ceeded by contraction and dilatation of the ventricles. There
is a short period, too, of complete relaxation or repose, as
if the heart was taking a momentary rest before entering
again on its duty.

The beat of the heart is felt between the fifth and sixth
ribs, near the breast-bone. The lower part of the heart is
called the apex, and is more movable than the upper part
or base, which is so attached as to be held firmly in place.
The position of the heart in the chest is well shown in
Fig. 55. The very smooth membrane covering the heart
being in two layers, the heart moves between them without
friction. A small amount of fluid is poured out on this
membrane to lubricate it. When the heart contracts it
alters its shape, and has a gliding motion against the walls
of the chest, to which the term *impulse* or *shock* has been
applied.

The movements of the heart are entirely beyond the
power of the will, although the muscles which make it up
appear to be like those which in other parts of the body
are under voluntary control. When the heart contracts it
sends out all the blood it contains, and when it dilates it
fills again with that fluid.

Vitality of the Heart.—The heart possesses a permanent power of contractility, which goes on through a long series of years without interruption. Although sometimes scarcely perceptible—after drowning, for example—the action of the heart may sometimes be restored by proper efforts at resuscitation. It is well worth while, therefore, under such circumstances to persevere for a length of time in our efforts to restore life; although there may be at first but a slight spark of vitality on which to build hopes of success. In some of the cold-blooded animals, the temperature of whose bodies is much below that of man, the heart will continue to beat for many hours or even days after its removal from the body. This is seen in the turtle, in some of the serpents, and in the alligator.

Sounds of the Heart.—When the ear is applied over the region of the heart, two sounds are heard—one, louder than the other, over the apex, the other over the base. The words *lupp, dupp* express the sounds heard. The first sound is also longer than the second. It is not certain whether this first sound is due to the closing of the valves, the movement of the blood through the openings, or the muscular contraction of the ventricles. It is not due to the shock of the heart against the chest, as it is heard in the heart of the dog after its removal from the body. The second sound seems to be connected with the closure of the semilunar valves. In its passage through the heart the blood takes the following course: The venous blood, when it comes back by the veins to the right side of the heart, enters the right auricle, which dilates and fills; the auricle then contracts and fills the right ventricle. This in turn contracts and sends the blood with some force into the pulmonary artery, which carries it to the lungs to be purified. After being aërated in the lungs the arterial blood returns to the left side of the heart, entering the left auricle. This cavity then contracts and sends the blood into the left ventricle, which by its contraction forces the blood into the arteries to be distributed through the body, and back by the veins to the right auricle, as before. The valves open freely so as to allow the blood to enter, but immediately close to prevent a reflow.

The Pulse.—The series of alternate contractions and dilatations is called the *beat* of the heart. When felt at the wrist or at any other superficial part of the body, as at the temple, it is called the *pulse*. These beats number in a healthy grown person an average of about seventy-two to the minute. The pulse is increased by food, by exercise, by heat, by rising from a horizontal to a vertical position, etc. Standing increases the number, giving the heart more labor to perform to send the blood to the head, etc., while lying down diminishes it. Fasting also diminishes it. The number of pulsations per minute—in other words, the frequency of beat of the heart—has relation to the quantity of blood in circulation. Thus we find the following facts in regard to the pulses of different animals:

	Quantity of Blood per 1000 grammes (= 15,000 grains).	Number of Pulsations.
Horse	152	55
Man	207	72
Dog	272	96
Rabbit	620	220
Guinea-pig	892	320

It will be noticed in this table that there is a steady increase of the number of beats in proportion to the amount of blood.

The effect of emotion on the frequency of beat of the heart is well known. Under unforeseen excitement it will beat violently, giving rise to palpitation. Sometimes under depressing influences, as grief or fear, the action of the heart may be so interfered with as to cause fainting, or even death from total suspension of its movements. These are illustrations of the effect of the nervous system on vital organs. The pulse may be naturally much slower or faster than the seventy-two beats mentioned as the average. Cases have been known in which through a lifetime the pulse has been as low as sixty, or even less, to the minute.

At birth the pulse of the infant is as high as one hundred and forty beats a minute, and during the early years of childhood it is much more rapid than in the adult. It again gains a few beats in old age. The pulse of the female is somewhat more rapid than that of the male—about ten beats faster—although the heart is in size slightly smaller.

The following is a near estimate of the average number of heart-beats at different ages:

At birth	140	Youth	90
Infancy	120	Adult age	72
Childhood	100	Extreme old age	75 to 80

The Heart's Life-Work.—The immense amount of mechanical work done by the heart during a lifetime may be estimated by the number of beats or pulsations, which, as we have said, indicates the number of times its cavities contract and dilate. Taking the number of pulsations as 72 to the minute, the following results are obtained: The heart beats 4320 times an hour, 103,680 times a day, or nearly 38,000,000 times a year. The new-born baby, with its numerous heart-beats per minute, in the first year of its life accomplishes more than 70,000,000 pulsations. During a lifetime of fifty years the heart will have beaten, at the least calculation, two thousand million times. Is it not

14

remarkable that this small organ, apparently left in charge of its own work, should be capable of such continuous and unceasing labor?

The Heart's Repose.—The only rest the heart gets is that momentary lull or repose which takes place after the second sound of the heart. This period seems to be of itself but slight and momentary, and yet at the end of the day, if we carefully calculate the amount, we find that the heart has had many hours of rest from labor. It has been estimated, indeed, that during twenty-four hours the ventricles work twelve hours and rest twelve, and the auricles work six and rest eighteen.

The Quantity of Blood.—If we estimate the quantity of blood sent out by the ventricles at each pulsation at four and a half ounces, the amount propelled from it during seventy-two pulsations, equivalent to a minute, would be 324 ounces, or about 20 pounds, being 1200 pounds an hour or nearly 13 tons a day. In all these estimates we seem to speak of the heart as if it were merely a self-sustaining machine, forgetting, apparently, for the moment, that its vital powers are inherent in it, and cannot be absolutely explained on the principles of any mere mechanical apparatus of human construction.

The whole quantity of blood in the body has been estimated as being in man $\frac{1}{14}$ of the weight of the body; in the dog, $\frac{1}{13}$; in the cat, $\frac{1}{15}$; in birds, $\frac{1}{12}$; in fishes, $\frac{1}{60}$.

The Arteries.—When the blood passes from the right ventricle into the pulmonary artery, or from the left ventricle into the aorta—the largest artery in the body—it is prevented, as already stated, in each case from flowing backward by valves, called from their shape *semilunar* (" half-moon "), which come together and completely shut up the cavity (Fig. 58, 7, 8). Here begins, at the aorta, the system of vessels called the arteries, which, dividing and subdividing, carry the blood which has been purified in the lungs, everywhere throughout the body for purposes of nutrition (Fig. 60).

The arteries are solid elastic tubes composed of three different coats, the middle one of which is muscular and elastic in the larger vessels, and decidedly muscular in the minuter arteries. It was supposed by the ancients, in their ignorance of the nature of the circulation, that they contained air; hence their name (from two Greek words meaning "to contain air"). The arteries are of a tough structure, to bear the heavy pressure of the blood when sent into them by the force of the heart, and are lined by a very smooth membrane, so that the circulation may go on evenly and without interruption through the elasticity and contraction of these vessels. After death they are always found empty, the blood accumulating in the venous system.

FIG. 60.—AN ARTERY.

The arteries divide and subdivide until at last the minutest vessels are only perceptible by means of a microscope, and so numerous are the branches sent out that there is hardly a portion of the body that is not thoroughly permeated with them. The beat of the heart, communicated to the arteries, which constitutes the pulse, is felt at the radial artery at the wrist only because the artery is there near the surface. By a wise arrangement, almost all the other large arteries are more deeply seated, beyond the risk of injury, protected by muscles, bones, etc. The pulse is felt by the physician chiefly that he may learn something of the force and frequency of the heart's action, and its reg-

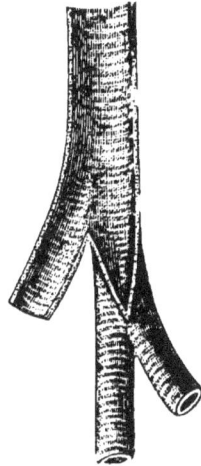

ularity and fulness, for these are guides to his knowledge of the healthy condition of the individual, the heart being itself affected or moved by sympathy with diseased conditions existing in other organs.

Pulse-Writing.—An instrument, called a *sphyg'mograph* (from two Greek words signifying " pulse-writer "), has been devised, so that the pulse can write on paper the line of its own travel. It is a long lever, moved by a screw acting on a small horizontal wheel, the point of the screw resting on a flat disk of ivory which rests on the pulse.

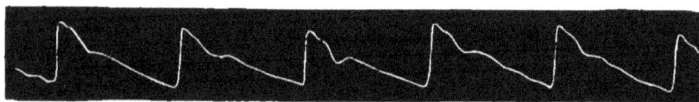

FIG. 61.—TRACING OF THE PULSE AT THE WRIST IN HEALTH.

The movement of the lever, carried along by clockwork over a blackened surface, gives the tracing (Fig. 61). Of course in disease the regularity and direction of this line would be varied, and the study of the causes of the variation would give the physician much information for his guidance. When the ventricles contract the vertical line

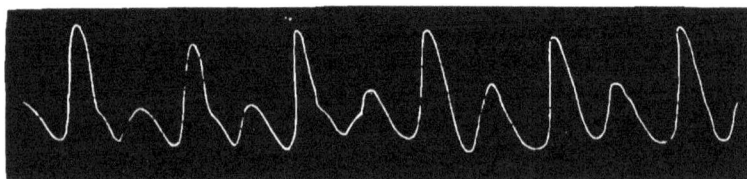

FIG. 62.—TRACING OF THE PULSE AT THE WRIST IN DISEASE, SHOWING A DOUBLE BEAT.

is recorded, and when they dilate they produce the wavy line. The effect of disease on the pulse is shown in one of its forms in Fig. 62, in which the tracing exhibits a double beat, an indication to the physician of serious mischief.

The Veins.—After the blood in the arteries has gone its

rounds by those vessels it returns to the heart by the veins. These are thinner and less elastic vessels, but are capable of distension. In health there is no pulse in a vein, and the blood flows continuously and not by pulsatile movement. When an artery is cut the blood flows from it in jets or spurts, on account of the great contractility of its coats; but not so with a vein, from which the blood "wells out" in a stream. When pulsation is felt in a vein it is probably due to obstruction of the circulation in the heart. The veins are minute at their origin in the different organs of the body, and gradually unite to form larger vessels, which communicate with each other, until they at last empty into the right auricle by two large trunks, the *venæ cavæ*, or hollow veins—one into the upper part of the auricle, the other (Fig. 57, *k, s*) into the lower.

Valves of the Veins.—The veins differ from the arteries in another important particular. They have *valves* (Fig. 63), or small membranous folds, arranged in such a way that they open to receive blood flowing in the direction of the heart, and become closed to prevent a return of the current in the opposite direction. They are more numerous where the blood proceeds against its gravity or where the parts around give them a feeble support. They are wanting in some of the important organs, such as the brain and the lungs.

FIG. 63.—VALVES OF A VEIN.

The veins are generally nearer the surface than the arteries, and contain a darker fluid. The safety accorded by nature to the arteries by placing them out of the way of injury in secluded places seems almost like an indication of intentional protection, by

which no harm shall result to the pure arterial blood in its duty of supplying nourishment to the body.

The Cap'illaries (from *capil'lus*, a hair, " hair-like ").— These are a system of very minute vessels, visible only under the microscope, and intermediate between the arteries and the veins. It is through the circulation in these vessels, which enter into the structure of every part of the body, that the nutrition of organs is effected. This capillary circulation may be seen under the microscope in transparent membranes, such as the web of a frog's foot, the wing of a bat, etc., in which we can trace the arteries and the capillary vessels coming off from them, finally merging into the veins. The more active or important an organ is, the richer it is in capillaries. The muscles, for instance, which are intended for constant movement, are fully supplied, while cartilages, whose action is more quiescent, are devoid of them. The capillaries are arranged in such a way as to be adapted to the organs in which they exist, being in network or meshes as seems best for their purposes. So completely is the system penetrated by these little vessels that the slightest scratch or cut may produce bleeding from the surface.

It would seem as if some of these vessels were at a great distance from the heart to be affected by its action, but the motion of the blood in the capillaries is doubtless due partly to that influence as well as to the natural elasticity of the vessels. The presence of capillaries in various parts of the body is readily shown by accidental causes, such as a grain of sand inducing a bloodshot eye. Blushing is an illustration also of the existence of capillaries, which under emotion become filled with blood.

Velocity of the Circulation.—The velocity of the circulation is affected by friction, by gravity, by curves in the course of the vessels, by division into branches, etc. If gravity interferes with the rapidity of circulation in one direction, it must aid it in another, as in parts of the body in which the blood flows quickly downward in the arteries and more slowly upward in the veins. The physician bears this in mind in the posture in which he places his patient, his head being elevated if there be any inflammatory condition of that part, the leg elevated in affections of the foot, etc. By a wise provision, the main artery of supply of blood to the brain takes a curved and tortuous course rather than a straight direction to that organ, so that the force of the current may be broken before it reaches its delicate structure.

Experiments have been made on animals to determine the rapidity with which substances introduced into the blood will return to the same point after having traversed the whole course of the circulation. These have proved that in most animals the rapidity of the circulation is equal to the time in which the heart makes twenty-seven pulsations. If this be true of man, and his pulsations be seventy-two per minute, the blood would occupy about $\frac{27}{72}$ of a minute in passing successively through the heart, the capillary vessels of the lungs, the arteries, the general capillary system, and the veins, or twenty-two and a half seconds. This is surprisingly rapid, considering the delicate structures that the blood must necessarily permeate, and the amount of nutriment and other material it is constantly giving off in its course.

Infusion and Transfusion.—By *infusion* is meant the introduction into the blood of various materials, such as medicines ; by *transfusion*, the introduction of fresh blood from one animal into the vessels of another. Both operations were first performed in the seventeenth century. As it was supposed that the cause of all diseases resided in the blood, it was imagined that all that was necessary was to replace the diseased fluid with pure, healthy blood. This process, which we call transfusion, is resorted to in cases of hemorrhage particularly, and in very low conditions of the system, when some extreme measure of this kind seemed absolutely necessary, has sometimes restored the patient to life.

When medicines are introduced by infusion into the bloodvessels they act more rapidly than when taken by the mouth, and produce just the same effects. An emetic, for instance, will vomit rapidly in this way. When transfusion was first used, soon after the discovery of the circulation by Harvey, it was thought that the operation would give young blood to old persons, and restore the vigor and vitality of their earlier manhood, and it soon became a popular remedy. It is at the present day resorted to only in cases of utter prostration, as already mentioned, for the restoration of life.

Circulation in Other Animals.—The circulation in the higher classes of animals—the mammalia, as they are called—is like that of man. The heart is composed of two distinct parts, each of which has its auricle and ventricle. In each case there is a double circulation, the greater and lesser, as already described. The situation of the heart differs, however, in some of them; and in the pig and the stag there are two small flat bones, called bones of the heart, which are found where the aorta leaves the left ventricle. Other differences also occur, as with the otter, dolphin, etc., some of the vessels of which are very large and tortuous, probably to act as a receptacle for the returning blood, while respiration is temporarily suspended during the existence of the animal under water.

In *birds*, the course of the circulation is very much the same as in man. The blood reaches the heart and is distributed from it by arteries, veins, auricles, and ventricles. In the right ventricle, however, the valve is sometimes replaced by a strong triangular muscle, which helps to force the blood from the right ventricle to the lungs. The lungs of a bird do not expand like those of man or the upper classes of animals, and therefore need this assistance.

In the *circulation of reptiles*—as the tortoise and lizard (Figs. 64, 65)—there is but one ventricle, instead of two, and there is a direct communication between the arterial and venous blood. Only a portion of the latter becomes purified in the lungs and mixes in the ventricle with the venous blood. There are usually two auricles—the left receiving the pure blood, the right the venous blood—both auricles discharging their contents into the common ventricle. It will be seen that the blood is but imperfectly arterialized, and it will be readily understood how, life being maintained with such imperfect aëration of the blood, the circulation may continue when the animal is so placed as to be incapable of respiration.

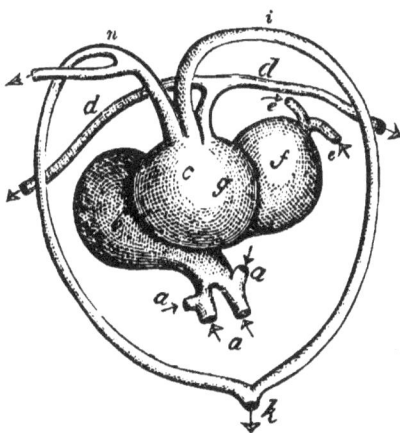

FIG. 64.—CIRCULATION OF THE TORTOISE.

a, a, venæ cavæ; *b*, right auricle; *c, g*, right and left ventricles united in one; *d, d*, pulmonary arteries; *e, e*, pulmonary veins; *f*, left auricle; *n*, aorta; *i*, large artery—really a second aorta—uniting at *k* with the original aorta.

The *circulation in fishes* is very simple. The heart has but one auricle and one ventricle. The blood is propelled from the ventricle to the gills, where it is arterialized, as has already been stated under Respiration (p. 143). Having thus acquired its purified state by contact with oxygen, it passes into a large artery analogous to the aorta, and thence into the general circulation, but does not return to the heart until it has passed through the capillaries. It enters the auricle and then goes to the ventricle, and thus the circulation is complete. The heart is extremely small in proportion to the body (Fig. 66).

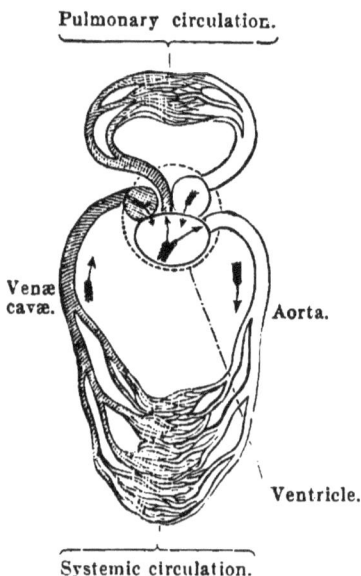

The *circulation in insects* is much more imperfect. They have neither arteries nor veins. The fluid which nourishes them seems to diffuse itself through their tissues. There is no heart, but a membranous tube runs along the back, in which alternate dilatations and contractions are perceptible, and this has been considered by some to represent a heart. The blood is watery and without color. In some of them there is a kind of movable valve, which gives a rapid movement to the circulating fluid. Insects differ, however, so much in their anatomy that no one description will apply to a sketch of the circulation in all. Some of the more highly organized seem to have bloodvessels, and in worms there is something very like a heart. The leech has no heart, but circulating vessels which dilate and contract.

In other classes of animals, such as spiders, lobsters,

Pulmonary circulation.

Venæ cavæ.

Aorta.

Ventricle.

Systemic circulation.

FIG. 65.—THEORY OF CIRCULATION IN REPTILES.

crabs, snails, etc. (Fig. 67), there are other peculiarities of structure for purposes of a circulation. Some of the spiders have a more perfect circulation than insects. The blood is white, but in the more highly organized there is an elongated heart at the back, which sends out blood and receives it again through the lungs. In the lobster and other shellfish there is a heart with one ventricle—but no auricle—and bloodvessels, and also something like lungs. The veins constitute a sort of reservoir rather than regular vessels. In snails, oysters, and animals of that class, the heart is composed of a single ventricle, from which the arteries pass, while the auricles are either double or single, and receive the red blood from vessels that seem to resemble the pulmonary veins, which in man carry the purified blood from the lungs. In the very lowest forms of animal life the blood or fluid which nourishes it is diffused by a sort of permeation or infiltration through the wall of the digestive tube, in the absence of either heart or vessels.

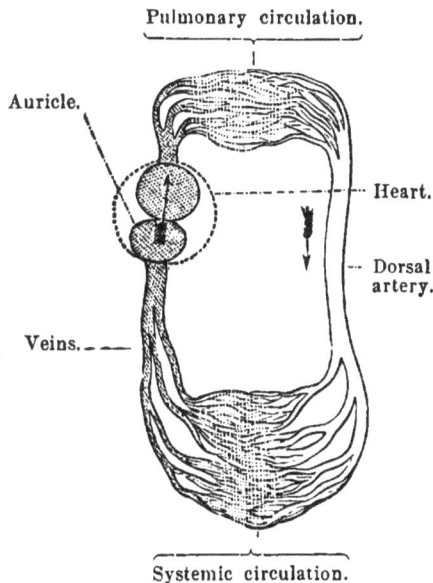

The Blood.—Having now described the apparatus by which the blood is distributed to the various organs for purposes of nutrition, we naturally turn to consider the general properties of this fluid which adapt it for the per-

Fig. 66.—THEORY OF CIRCULATION IN FISHES.

formance of its duties. It is estimated that the entire amount of blood contained in the body of a grown person is about one-fourteenth of his whole weight. To enable it to maintain the nourishment of the body, the materials which enter into its composition should be the same as those which make up the body itself. This we find to be the case. In some of its properties it resembles the chyle, which we have already seen was formed as a result of the digestive process, and which is absorbed into the chyliferous vessels in the intestines to become blood. It of course differs in color, but chyle and lymph are both rudimental blood. At first, when drawn from the arm, as in bleeding a patient—a practice that was formerly quite popular—the blood does not seem to be the mixed fluid that it really is. It is of a dark color as it flows from the veins, and a bright red or scarlet in the arteries. It is about 100° in temperature in man, but in some other animals much higher, as in the sheep, in which the temperature is 107°. It is heavier than water.

Pulmonary circulation.

Heart.

Arteries and capillaries.

Systemic circulation.

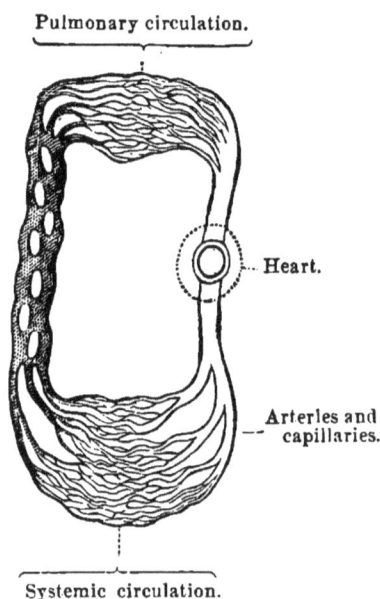

Fig. 67.—Theory of Circulation in Crabs, Lobsters, etc.

The Blood-Globules.—If it were not for the microscope we should not know of what parts blood is composed. When submitted to examination by that instrument it is found to consist of a large number of minute red particles—microscopic bodies called blood-cor'puscles, blood-globules, or discs—suspended in a thin fluid called the serum. These bodies are so minute that the smallest drop of blood on the point of a needle contains myriads of them. They are of a regular and definite shape in the same animal, but are of different form in different animals. In the class of animals called mammalia, which includes man and the animals next in the scale, the cor-

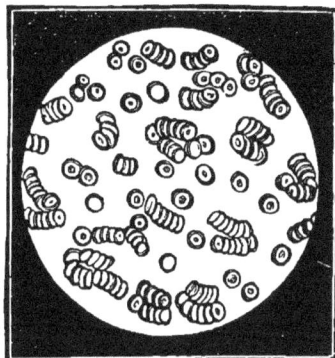

FIG. 68.—HUMAN BLOOD-COR-PUSCLES (magnified).

puscles are circular, while in birds and cold-blooded animals they are elliptical (Fig. 69). The human blood-cor-

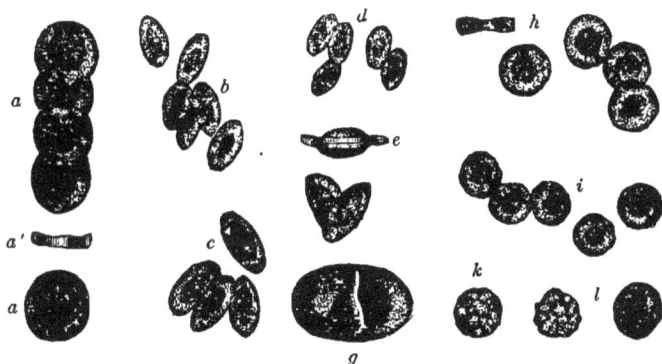

FIG. 69.—BLOOD-CORPUSCLES OF VARIOUS ANIMALS (greatly magnified).

a, a', blood-globules of man, seen under different views; b, of the camel; c, d, of birds; e, of the frog, seen edgewise; f, of the proteus; g, of the salamander, the external membrane being stripped; h, of the lamprey; i, of the lobster; k, of the slug-snail; l, two white globules of human blood.

puscles are flat discs, somewhat concave in the middle, with slightly rounded edges (Fig. 68).

Some idea of their minuteness may be obtained from the

15

statement that in man their average diameter is only the $\frac{1}{3200}$th of an inch, and their thickness the $\frac{1}{12000}$th of an inch. The serum in which they float is transparent and colorless, the redness being entirely due to the red corpuscles. They do not appear to be so red under the micro-scope as in reality, but this is an optical effect produced by the thinness of the medium in which they are observed by the eye. Sometimes the corpuscles arrange themselves in rolls—or rouleaux as they are called, like coin piled up together—and then they seem of a redder color to the eye. The size of the corpuscles varies in different animals. We can readily understand how important it is to be able to distinguish under the microscope the difference between the corpuscles of man and other animals. In a case of supposed murder the life of one suspected of the crime might depend on the result of such an examination. It is not many years since the blood of a pig on the clothing of a supposed murderer placed the life of the prisoner in imminent peril.

To show more clearly the varying size of the corpuscles in animals as compared with man, the following brief statement may be made, the measurement being first given of animals with circular corpuscles or discs, and afterward of those with elliptical corpuscles.

Of animals with circular discs, a few examples will suffice :

	Diameter.
Man	$\frac{1}{3200}$ inch.
Elephant	$\frac{1}{2745}$ "
Musk-deer	$\frac{1}{6500}$ "

In other animals which have elliptical blood-corpuscles the measurement must be taken both in length and width :

	Long Diameter.	Short Diameter.
Camel........	$\frac{1}{3250}$	$\frac{1}{3921}$
Ostrich...............................	$\frac{1}{1650}$	$\frac{1}{3000}$
Pigeon.................................	$\frac{1}{2314}$	$\frac{1}{3425}$
Humming-bird........................	$\frac{1}{2666}$	$\frac{1}{4000}$
Frog	$\frac{1}{1108}$	$\frac{1}{1621}$
Crocodile	$\frac{1}{1231}$	$\frac{1}{2286}$
Shark..................................	$\frac{1}{1143}$	$\frac{1}{1684}$
Earth-worm	$\frac{1}{110}$	$\frac{1}{1200}$

White Corpuscles.—White corpuscles are also found in the blood, of a spherical shape, and not so well defined as the red corpuscles, although somewhat larger in size. They are much fewer in number than the red, and have a more sluggish movement in the vessels. In healthy blood there is about one white or colorless corpuscle to every four hundred or five hundred red corpuscles. These are considered to be the globules absorbed into the blood from the lymphatic vessels and the chyle, and to become developed afterward into the red corpuscles.

Number and Uses of the Red Corpuscles.—The number varies greatly in different animals, seeming to bear a pretty close ratio to the temperature. In all of them they are to be counted by millions. One estimate, generally considered accurate, is that five millions of them are contained in

FIG. 70.—BLOOD-CRYSTALS.

the space occupied by a very small drop of blood. Another authority has stated that if the colored corpuscles of the adult man were placed side by side on a flat surface they would cover an area of about 3000 square yards. The higher the temperature the greater the number of the corpuscles. By chemical action beautiful crystals can be obtained from the coloring-matter of the blood, which under the microscope present the appearance presented in Fig. 70. The blood-globules are really carriers of

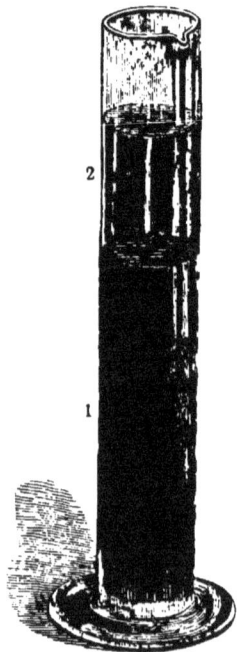

oxygen, which they obtain in the lungs during respiration, as has been already shown, and bear with them to the different tissues. Parts which are in active exercise, such as the muscles and nerves, need this fresh supply of oxygen, and in the course of the wear and tear to which they are subjected give to the blood, by lymphatic absorption, a certain amount of carbonic acid, which passes along with the blood, to be gotten rid of in the lungs by expiration. It will be readily seen why hemorrhages are fatal or serious in character. Life can only be maintained as long as the blood-corpuscles are well organized and contain the proper proportion of oxygen. An animal cannot survive the loss of any large quantity of its blood. A certain amount of iron exists in the blood

FIG. 71.—COAGULATION
OF THE BLOOD.
1, clot; 2, serum.

but not in any great proportion, probably not more than thirty grains in the whole body.

Coagulation of the Blood.—Blood that is circulating in the body consists of two portions, the red corpuscles and a watery portion called the liquor sanguinis (" water or solution of blood "). When blood is drawn from the body a rapid change takes place in it. Instead of a homogeneous fluid, such as it is in the vessels, it becomes separated in a short time into distinct portions—a reddish jelly-like, trembling mass, to which the name *clot* is applied, and a yellowish liquid called the *serum*. The cause of this change is due to the separation from the blood of an element in it called fibrin, which is soluble in living blood —if we may so call it—and insoluble in dead blood, or that which has been drawn from the bloodvessel. The fibrin leaves the liquor sanguinis, of which it formed a part, and draws down with it to the bottom of the cup or receptacle in which it has been received the red corpuscles. The clot is therefore the union of the red globules and the fibrin of the blood; the serum is the thin liquid portion left behind, in which are dissolved the other ingredients of the blood (Fig. 71).

The difference between circulating and coagulated blood may be clearly exhibited in the following diagram :

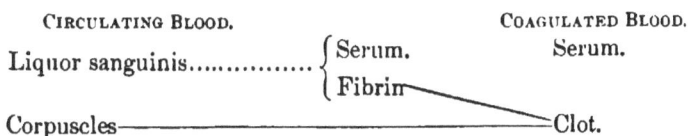

CIRCULATING BLOOD.   COAGULATED BLOOD.

Liquor sanguinis............... { Serum.   Serum.

{ Fibrin

Corpuscles————————————————Clot.

(The lines indicate the combination of the fibrin with the corpuscles to form the clot, while the serum is left alone.)

Various attempts have been made to explain the causes of coagulation of the blood when exposed to the air, but this is one of those difficult points which do not admit of ready explanation. Some physiologists content themselves with calling it a " vital " process, which is a quiet admission that they know nothing of its causes beyond the fact that it is in some way connected with the life of the blood or of the individual, and is not merely a physical process. All we know is, that there is in the blood an element called fibrin which becomes entangled with the corpuscles to form the clot.

15 *

Whatever its exciting cause, coagulation is an important means of stopping bleeding after a bloodvessel is cut by accident or otherwise, the clots formed stopping up the open mouth of the vessel. In some animals, as birds, coagulation, or the formation of a clot, takes place instantaneously. Thus Nature seems constantly to guard the lives of some of the most helpless of her creatures by placing within their frames the elements of preservation of life itself, even when danger most seriously threatens it. It may be asked why the blood does not coagulate in the vessels during life when exposed, as it always is, to friction and motion. It does sometimes, but rarely, form clots, which block up the smaller vessels, and possibly a cavity of the heart, and endanger life, but while circulating in living tissues and brought in contact with them coagulation is of the rarest possible occurrence.

QUESTIONS.

Through what fluid are all parts of the body nourished?

What is the process called by which the blood is sent through the body?

What relation has it to digestion, absorption, and respiration?

To which side of the heart does the blood go after leaving the lungs?

What is the great central organ of the circulation?

Why is the process called a circulation?

Of what tissue is the heart chiefly composed?

What is the shape of the heart? What is its position in the chest?

What is its weight? Its size?

What is the lining membrane of the heart called?

What does the word endocardium mean?

How is the heart covered?

What does the word pericardium mean?

How many cavities are there in the interior of the heart?

How many of these propel the blood?

Is the heart a single organ?

Which side of the heart receives the venous blood?

To what organs does the venous blood pass after leaving the right side of the heart?

Where does the blood go after leaving the lungs?

What are the cavities of the heart called?

Trace the course of the blood from the right auricle to the left ventricle.

Into what set of vessels does the blood pass after leaving the left ventricle?

What are arteries and veins?

What is the venous heart? The pulmonary heart? The arterial heart? The aortic heart? The systemic heart?

What is the lesser or pulmonic circulation?

What is the greater or systemic circulation?

What is the object of each circulation?

Which side of the heart has the most labor to perform?

What arrangement is there to prevent the blood from flowing backward in its course?

What is the valve between the right auricle and right ventricle called? Why?

What is the valve between the left auricle and left ventricle called?

Who first discovered the circulation of the blood?

What are the semilunar valves?

To what are the regular movements of the heart due?

What is the systole? The diastole?

What is the nature of the movement of the auricles and ventricles?

Does the heart have any rest from labor?

Where is the beat of the heart felt?

What is the apex of the heart?

How is the movement of the heart facilitated?

What is the impulse of the heart?

Is the heart made up of voluntary or involuntary muscular fibres? Is it under control of the will?

What effect on the blood in it has the contraction of the heart?

Is there any limit to the contractile power of the heart?

What heart-sounds are heard when the ear is applied over the chest?

Which sound is the longest?

What is the cause of the first sound?

Is it due to the impulse of the heart against the chest?

What is the cause of the second sound?

Describe the course of the blood through the auricles and ventricles.

What is meant by the beat of the heart?

What is the pulse?

How many times does the pulse beat in a minute?

How may this number be increased?

What effect has position or attitude on the pulse?

Has the pulse any relation to the quantity of blood?

What varying effects have excitement, emotion, etc. on the pulse?

At what period of life is the pulse most rapid?

In which sex is the pulse more rapid?

What effect has sex on the size of the heart?

How often does the heart beat in a minute? In a day? In a year?

How much blood is sent out from the ventricles at each pulsation? How many ounces per minute?

What proportion of the weight of the body is the blood?

How does this compare with other animals?

What action have the semilunar valves of the heart?

What is the duty of the arteries?

What is the largest artery in the body?

Why were arteries so named?

What coats have the arteries?

What is the condition of the arteries after death?

What is the mode of division of the arteries?

Why do we feel the pulse at the wrist? What do we learn from the pulse?

How are the arteries protected in various parts of the body?

What is a sphygmograph? What is its importance?

What are the veins? How does a vein differ from an artery?

What is the difference between the flow of blood from a vein and an artery when cut?

Is there any pulse in a vein?

What are the venæ cavæ?

Do arteries or veins possess valves? For what purpose?

Are valves present in all the veins?

What are the capillaries?

Through what vessels is nutrition effected?

What organs are most richly supplied with capillaries?

What keeps up the movement of the blood in the capillaries?

What vessels are interested in blushing or in a bloodshot eye?

What effect has gravity on the rapidity of the circulation?

How is the force of the current of blood to the brain checked?

What do we learn from experiment as to the velocity of the circulation?

In what length of time does the blood pass through the body?

What is meant by infusion? By transfusion?

Under what circumstances are these operations resorted to?

What is the nature of the circulation in the higher classes of animals?

In what respects do the heart or vessels of the stag, dolphin, etc. differ from those of man?

What is the peculiarity of the arrangement for circulation in birds?

How many ventricles has the tortoise?

Is there any direct communication in them between the venous and arterial blood?

What effect has this upon the purity of the blood?

How many auricles and ventricles have fishes? How, then, is the blood arterialized?

What becomes of the blood in fishes after it leaves the gills?

What organs are absent in some insects? What is the color of their blood?

How do insects themselves differ in their circulation?

What is the nature of the circulation in spiders? In shell-fish? In snails?

How does the blood circulate in the very lowest forms of animal life?

What proportion does the amount of blood in the body bear to the whole weight?

What fluid formed in digestion resembles the blood?

What is the color of the blood in the veins? In the arteries?

What is the temperature of the blood in man? In other animals?

How does its weight compare with that of water?

Of what parts is the blood composed as seen under the microscope?

What is the shape of the blood-globules in man and the higher animals? In birds?

What is the size of the human blood-corpuscles?

What is the importance of distinguishing between the shapes of the globules of man and of other animals?

What other corpuscles are found in the blood?

State from the table some of the animals that have the largest blood corpuscles.

How do the shape and size of the white corpuscles compare with those of the red corpuscles?

How many red corpuscles are there to every white corpuscle?

What are the uses of white corpuscles?

What is said as to the number of the red corpuscles in the body?

What are the uses of the red globules of the blood?

Where do they obtain oxygen?

M

Where does the carbonic acid in the blood come from? How is it got rid of?

How much iron is present in the blood?

What change takes place in the blood when drawn from the body?

What are the two portions called?

What element separates from the blood to help form a clot?

What is the clot composed of?

What is the difference in composition of circulating blood and coagulated blood?

What effect has coagulation on bleeding?

Does the blood ever coagulate during life?

ANIMAL HEAT.

Temperature of Animals.—The temperature of the body characteristic of man and other animals is known as Animal Heat. It is in most animals a constant quantity; that is, each animal has a temperature which does not usually vary. Such animals are known as *warm-blooded*, and the term *cold-blooded* has been assigned to those whose temperature is not constant and not much above that of the external air or the water in which they live. The temperature of the latter varies greatly also with conditions of the atmosphere, etc. surrounding them, and is considerably below that of the human body. Reptiles and fishes belong to the class of cold-blooded animals. The subject of Animal Heat is often taught in connection with that of Respiration, because the two processes are associated with or dependent upon somewhat similar chemical changes.

Sources of Animal Heat.—In the process of respiration more oxygen is taken into the lungs than is given off again in the form of carbonic acid, of which oxygen is an ingredient. This excess of oxygen serves a useful purpose all through the body by uniting with the carbon and hydrogen which have been taken as food, and this union gives rise to the production of heat. When carbon thus combines with oxygen, their union forms carbonic acid; and water is formed by the union of the oxygen and hydrogen. Every chemical change which occurs in the body results in

179

the production of heat. We notice a similar result occurring during the germination and flowering of plants.

Effect of Food, Respiration, etc.—The more active the
process of respiration and the more generous the supply
of food, the greater the amount of heat produced. Thus
animals, such as birds, which have a particularly active
process of respiration, evolve the greatest amount of animal
heat, as shown by the general temperature of their bodies.
Where the respiration is slow and inactive, as in reptiles,
and the same amount of oxygen is not required, the temperature is not so high. Increased exercise, which gives
rise to rapid breathing, elevates the temperature. In the
Arctic regions a larger quantity of food, especially of an
oily and fatty kind—which are composed of carbon and
hydrogen—is taken than by the dwellers in milder climates.
The influence of food upon the temperature of the body is
shown also in the fact that the animal heat is reduced during
starvation, and it is a well-recognized point that in such a
condition freezing to death takes place very rapidly.

Temperature of Different Organs.—The temperature of
the body may be ascertained in various regions. When a
thermometer is placed in the arm-pit, it registers in health
from 98° to 99° (Fahrenheit), and the temperature of the
body generally is stated to be 98½°. When the bulb of
the thermometer is placed in the interior of the mouth, as
beneath the tongue, the same degree of temperature is
recognized. Whatever it may be, it usually varies but a
degree or two. In young children the temperature is about
two degrees higher than in the adult. Of late years much
attention has been paid by the physician to the study of
changes of temperature in disease. Any decided variation
above or below the standard is a subject for anxiety. In
some affections, such as typhoid fever, there is an elevation

of temperature, while in cholera it may fall many degrees. In health any marked deviation is checked by the power of evaporation possessed by the skin and seen in the visible perspiration on the surface of the body.

The temperature of the various organs, as the lungs and the muscles, is rather higher than that of the surface of the body. The blood is hotter on the right side of the heart than on the left, and cooler after it leaves the heart, the temperature having doubtless been lowered in the lungs during the passage of the blood through them. The temperature is lower in the veins near the surface of the body than in the arteries, but in interior tissues and organs the blood coming from them by the veins is warmer than that going to them by the arteries. It has been thought by some that the heat of the body was produced by chemical action in the lungs during respiration, but this is not now considered to be the correct theory. If all the heat originated there, the delicate structure of the lungs would be undergoing a process of incessant combustion, which would soon destroy it.

Mechanical forces, such as friction, muscular movement, etc., also give rise to heat in varying quantities. The greatest amount of heat is produced in the liver and by the muscles, and it is found that the blood is warmer after coming from a muscle than it was in going to it.

Atmospheric Influences, Clothing, etc.—Man, having a temperature nearly constant under all circumstances in health, and being able to regulate the proper quantity of his food and exercise, as well as to defend himself against intense heat or cold by regulation of his clothing, can tolerate excessive heat or cold with much greater impunity than is possible with any other animal, whether warm-blooded or cold-blooded. No matter what the climate or

16

season, the temperature of the human body remains very nearly the same. The thicker clothing with which man protects himself in winter prevents the loss of the heat of the body by radiation.

The occupations of individuals sometimes render it neces sary for them to be subjected to very high temperatures, such as those of iron-works, etc., but even under such circumstances the temperature of the human body is not much affected. As already stated, there is a compensation for all this in the evaporation attending the increased amount of perspiration produced by such exposure. It is a well-known fact that during evaporation heat is abstracted, and the part becomes really cooler. The more rapid the evaporation the more decided the sensation of cold. In warming the rooms we occupy in winter we endeavor to prevent the heat of the body from being too rapidly lowered. The temperature of the body is always higher than that of any artificial heat we obtain or could bear in our residences. In cold seasons there is but a small amount of perspiration poured out, so that evaporation can produce but little coldness of the surface, and effect, if any, only a slight reduction of temperature of the body. We shall hereafter, as part of the study of the physiology of the touch, investigate the action of the skin while engaged in this important duty of exuding the perspiratory fluid.

We have thus traced the history and course of the blood from the time it was formed from food and lymph by Digestion and Absorption through its changes to a purer state in Respiration, and in its distribution through the body by the Circulation. The chemical changes involved in the production of Animal Heat take place through the agency of the capillary bloodvessels, which in myriads permeate even the minutest portions of the body.

QUESTIONS.

What is meant by Animal Heat?
Does the temperature vary much in the same animal?
What are warm-blooded animals?
What are cold-blooded animals?
Give examples of the latter.
Is more or less oxygen taken into the lungs than is given off from them?
What purpose does this excess of oxygen serve in the system?
What is formed when carbon and oxygen thus unite? When hydrogen and oxygen unite?
By what kind of action is the heat of the body produced?
Does this occur in the vegetable?
What effect have food and respiration in the production of heat?
How is this exemplified in birds?
When respiration is sluggish, how is the animal heat affected?
What effect has exercise?
In the Arctic regions what is the relation of food to animal heat?
What effect has starvation?
What is the usual temperature of the human body? Of the region under the tongue?
How does the temperature of young children vary?
What information does the physician get from the temperature?
What effect has the skin in checking elevation of animal temperature?
Is the external or internal temperature the higher?
On what side of the heart is the blood the hotter?
Where is the blood of the heart cooled?
Is the temperature lower in the veins or the arteries?
In what organ was all the animal heat at one time supposed to be formed?
What mechanical forces produce animal heat?
In what organs is the greatest amount of heat produced?
Why is man able to tolerate excessive heat or cold?
What is the effect on animal heat of his thicker winter clothing?
What effect has evaporation on the heat of the body?
How is the amount of perspiration and evaporation affected in cold weather?
What bloodvessels are concerned in the production of animal heat?

SECRETION.

Materials Separated from the Blood.—While the blood is going the round of the circulation, supplying nourishment at every point, the cells to which we referred in the introductory chapter (p. 20) are busy taking from the blood various matters, which, being of no use in the system, must be partially or entirely separated from it. If it were not for some such arrangement as this, these useless materials would accumulate everywhere to the injury of the health, and would clog up the vessels so that the individual would surely die. This is the reason that such channels as the tears and the perspiration exist for getting rid of matters that are not nutritious. We have already stated that the cells everywhere are endowed with the power of selecting from the blood a special kind of material, just as if they had a mind of their own to guide them. Those cells, for example, which take materials from the blood with which to form tears always choose the same kind of materials, and never form anything else except tears, and those which form the fluid of the perspiration never create anything else than perspiration.

Secretion only Separation.—In referring to digestion in the intestine, mention was made of the pancreatic juice and the bile, and the salivary glands were also alluded to in connection with the part performed by the mouth in digestion. These are all specimens of *secretions*, so called ; that is, of fluids separated from the blood by the action of

184

cells, for secretion means nothing more than separation. If there is any great amount of fluid separated in this way. there is a large organ or apparatus for the purpose. The bile, for instance, is secreted by the liver, which is one of the most important organs in the body. The simplest forms of separation of matters from the blood occur by a process like filtering from the delicate bloodvessels through a membrane ; and this is aided by the pressure of the blood itself.

Glands.—The organ or body whose duty it is to separate materials from the blood is generally called a *gland.* Thus we have glands of the skin, which take matters from the blood to form the perspiration. The liver, pancreas, etc. are also examples of glands. They always have a canal or duct leading from them to carry off the fluid which they have been forming, and to empty it somewhere. The bile, for instance, is poured out by a canal (p. 104) into the intestine ; the perspiration is poured over the surface of the skin through thousands and thousands of little tubes or canals leading from the sweat-glands. Sometimes, as in this case, there are many of these little tubes ; the gland over the eye, which forms the tears, being also another example of the same tubular plan.

Various Kinds of Secreting Surfaces.—The simplest arrangement for purposes of secretion is an animal membrane and a minute bloodvessel (Fig. 72, *b, c*). In other cases, as in that of more complicated organs, like the liver, the action of cells is necessary in addition to the membrane and vessel, because the power of selection is required. As the skin or any other form of membrane would not of itself give surface enough for the immense amount of work it has to do in separating materials from the blood, it is absolutely essential that the extent of surface should be in-

16 *

creased in some way. This is done by one or more of the
arrangements shown in Fig. 72, in which, by a turning in,
or involu'tion, as it is sometimes called, of the secreting
membrane, room is given for a very large distribution of
vessels, cells, and membrane, all of which are necessary
to perfect secretion. A single inch of skin in this way
becomes prolonged into a surface of many yards. This is

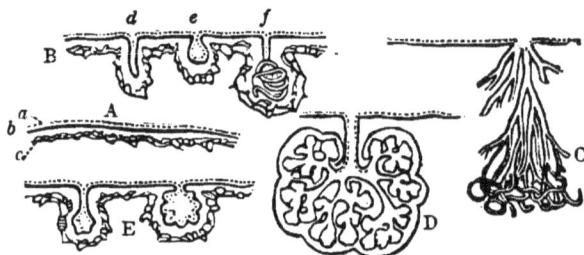

FIG. 72.—ARRANGEMENT OF SECRETING STRUCTURES.

A, a, cells; b, membrane; c, capillary bloodvessels; B, simple glands, showing their
different kinds of secreting apparatus; d, straight tube; e, sac; f, coil of tubes;
C, compound tubular gland; D, gland arranged like bunch of grapes; E, other
shapes of glands. The dotted lines represent the layer of cells through the
agency of which secretion is effected.

not at all visible to the naked eye, and can only be seen
under the microscope. Many thousands of such involu-
tions of the surface and of such arrangements for secretion
would thus occupy but a small amount of space. It will
be noticed that some of the glands are tubular—that is,
they are made up of tubes; others are in shape like a
bunch of grapes (Fig. 23), while others again are mere
sacs.

In the lower back part of the abdomen are two of the
most important glands of the body—the kidneys—one on
each side of the spinal column. A large amount of blood
is sent to them, and from this blood useless solid and liquid
materials are separated, to form a fluid which afterward
passes, drop by drop, into a sac called the bladder.

QUESTIONS.

What kinds of materials are separated from the blood?

Why is it necessary that these matters shall be separated?

What power have cells in selecting materials from the blood?

What kind of secretions are poured into the intestine during digestion? Into the mouth?

What is meant by a secretion?

How is the simplest form of secretion effected?

What is a gland?

What are the principal glands found in the skin?

How is the secretion of a gland poured out or emptied?

How are the bile and the perspiration poured out?

What two parts are necessary in the simplest arrangement for secretion?

What other action is necessary in the larger glands?

What, then, are the three parts usually necessary to effect perfect secretion?

How is the extent of surface for secretion increased?

What are some of the shapes of the glands when seen under the microscope?

THE NERVOUS SYSTEM.

A Function of Animals Alone.—A Nervous System is peculiar to animal life, having no existence in the vegetable. In this lies the one great feature distinguishing vegetables from animals. We have already seen that digestion, absorption, respiration, and circulation take place in both animals and vegetables. By means of a nervous system the animal is placed in direct communication with the world around it, and by it the different organs are able to act harmoniously together. It is developed in proportion to the scale of intelligence, and man therefore possesses a nervous organization which gives him a superior position over the whole animal kingdom. By it man thinks and has sensation and voluntary motion. Even when unconscious, both men and animals are still controlled in their actions by a branch of the nervous system which is entirely beyond their power to control. Man lives although he may be asleep, and the nutrition of his body goes on whether waking or sleeping. It is very evident, therefore, that there must be a different nervous arrangement in charge of all such processes, and that there must be one portion entirely under his will, while another acts continuously and wholly beyond his control. In man, and all other animals which have a distinct nervous apparatus, we find it to be actually the case that there are divisions of the nervous system constituted in this way.

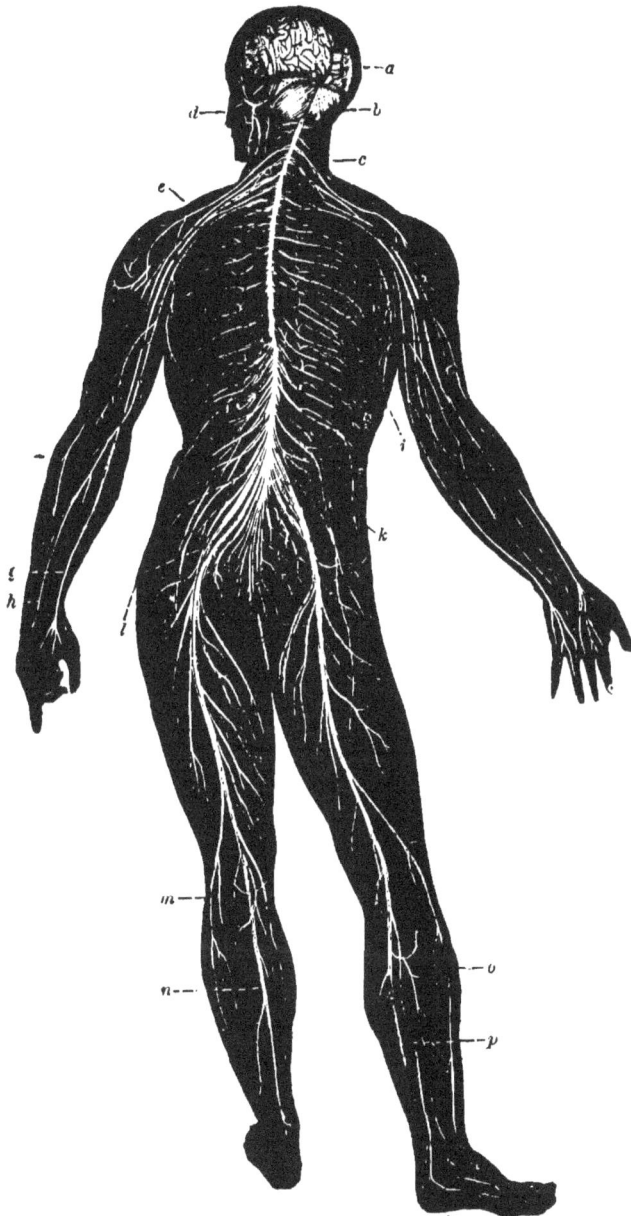

FIG. 73.—CEREBRO-SPINAL SYSTEM.

a, cerebrum; *b,* cerebellum; *c,* spinal cord; *d,* nerve of face; *e, f, g, h,* nerves of
arm; *i,* nerves between ribs; *k,* nerves of lower part of back; *l,* nerves in region
of hip; *m, n, o, p,* nerves of leg.

Divisions of the Nervous System.—One of these great nervous systems is regulated by the brain and spinal cord as centres, and is therefore called the *cere'bro-spi'nal system* (from cere'brum, " brain "), (Figs. 73, 75). Impressions are conveyed to and from these centres by slender white cords, called *nerves*, which divide and subdivide to extreme minuteness until every part of the body is supplied with them. No matter how minute a section or an organ of the body may be, little branches of nerves are found in it. There is not a point of the surface, for example, which does not, when cut or pricked with a pin, show by the pain and suffering experienced that it is fully supplied with nerves.

Another part of the nervous system is sometimes called the *sympathet'ic system*, because it brings all portions of the body into direct sympathy with one another, or the *ganglion'ic system*, because, instead of having a brain or a spinal cord to control it, it has numerous enlargements called ganglions or nervous masses, connected together by nerves so as to form one continuous chain (Fig. 84). This is a distinct nervous system, but its nerves have communication with the nerves of the other system in order to make the chain of sympathy between the various organs complete. We can readily understand the importance of this action of sympathy when we reflect that all the duties which the organs perform must go on, either continuously or as often as required by the needs of the system. For instance, the gastric juice is only excited under the influence of the presence of food, as already described under Digestion. This is an impression on the nerves of the stomach in the first place, through which the glands of that organ are stimulated to act. If it were not for a sympathetic nervous system, through which all the organs can communicate and sympathize with one another, each organ would be acting in its own independent course, and health and harmony would be impossible. We shall describe this great sympathetic system of nerves more fully hereafter (Fig. 84).

Nervous Matter.—This is a soft substance, almost fluid at birth and for a short time afterward. When viewed under a microscope it is found to be made up of a white and a gray matter. The nervous matter generally is composed of the white substance, which is arranged in delicate nerve-fibres about the $\frac{1}{6000}$ of an inch in diameter. The gray or ashy-colored nervous matter is composed of nerve-cells of various sizes and in large quantity. They are somewhat

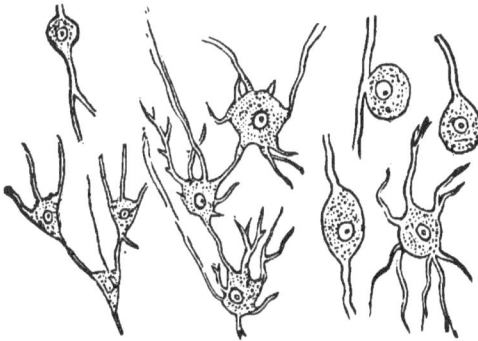

FIG. 74.—VARIOUS FORMS OF NERVE-CELLS.

rounded, with a nucleus or central spot on each, with elongations running off in various directions (Fig. 74). A ganglion is usually a collection of these nervous elements massed together, the gray matter being in excess. Some of the physiologists, who think the nerves begin at the surface of the body, consider that the brain and spinal cord are ganglia in combination with the white matter derived from the nerves. White fibres are largely found in the long cords called nerves, which pass to and from the various parts. The exterior of the brain is made up of gray nervous matter; the interior, white matter. The gray matter originates nervous power; the white conveys it.

The Cerebro - Spinal System. — The central parts of
the cerebro-spinal system
(Figs. 73, 75) are the
brain and the spinal cord,
both of which are soft
masses of white and gray
matter, variously arrang-
ed in these different or-
gans. Taking the mass
altogether as found in the
spine and in the skull, it
is called the *cerebro-spinal
axis.* The part contained in
the skull is the *brain;* that
in the spine is the *spinal
cord,* or *spinal marrow,* as
it is sometimes called.

All these parts of the
nervous system (Fig. 75)
are protected from injury
by bones and by mem-
branes covering them. For
instance, in the case of the
brain, a blow or a fall on
the head would affect that
delicate organ much less,
on account of the covering
of bones and membranes
with which it is protected,
than if the brain had been
nearer the surface, greatly
exposed to external injury.

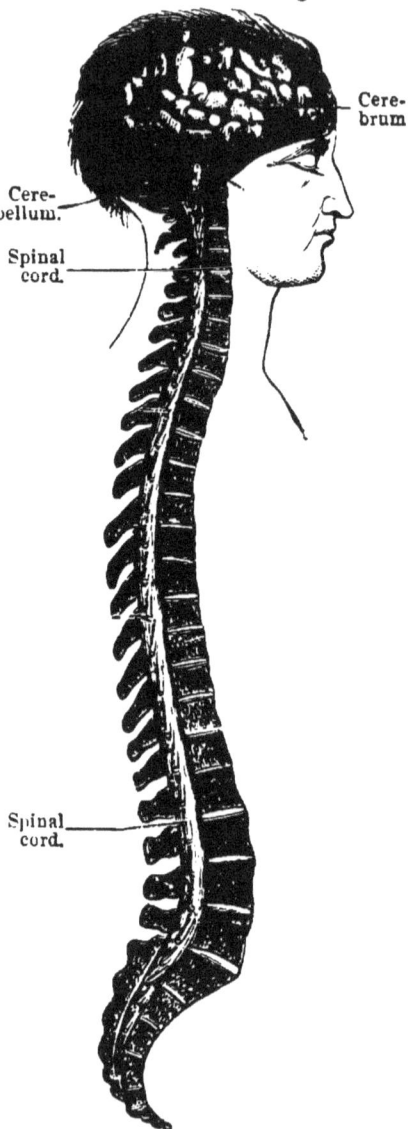

FIG. 75.—THE BRAIN AND SPINAL CORD.

The brain and spinal cord are
both covered with three coats, one within the other, the

outer coat being thick and strong, adhering in some places to the bones, and dipping down between folds of the brain to form a kind of partition for its protection from pressure from the other hemisphere or from the cerebellum.

One of these coats is like a spider's web, so thin and delicate is its structure. Between this membrane and the next one is a space filled with a liquid, called the *cere'bro-spi'nal fluid*. It seems to have been placed there to prevent the surfaces of the brain from rubbing against one another, just as oil is poured on parts of machinery that come into contact with other parts. If it were not for this fluid, the head when moved on the spinal column, when bent in walking or stooping, would press on the delicate nervous matter and injure it. In the middle coat are thousands of little bloodvessels running in all directions. The membrane not only itself protects these small vessels, but acts like a breakwater in diminishing the force of the current, which would otherwise injure the brain.

The Brain (Figs. 73, 75, 76, 79).—This wonderful organ, placed in the skull, is really divisible into three distinct parts, all associated together. These are the brain, or *cere'-brum*, the large round mass which fills the upper, middle, and front portions of the skull ; the little brain, or *cerebel'-lum*, a smaller, flattened portion at the lower and back part of the skull ; and a still smaller part, the *medul'la oblonga'ta*, which, translated into English, means the spinal cord prolonged (into the skull), for it is only the extension upward of that important portion of the nervous system. There is a large opening in the under and back part of the skull, which is placed there for the very purpose of admitting this organ to pass through it. There is a part of the brain which acts as a bridge between the cerebellum, or little brain, and the medulla oblongata, and this has been called the *pons Varo'lii,* or bridge of Varolius, an anatomist of Italy, who first described it.

If we should measure the amount of space in the skull occupied by the whole brain, we would find that seven-eighths of it belonged to the cerebrum, or brain proper. This part of the brain is the seat of the intellect or intelli-

Convolutions.

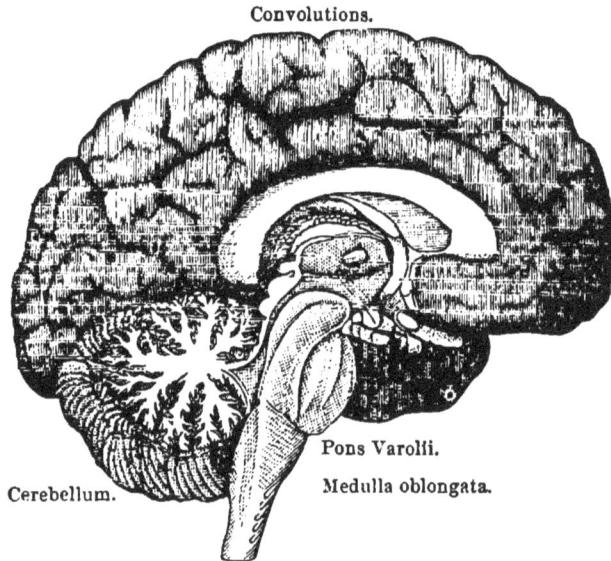

Pons Varolii.

Cerebellum.

Medulla oblongata.

FIG. 76.—INTERIOR OF THE BRAIN—showing, by section through it, its various parts.

gence of the animal, and man has a greater amount of it, in proportion to the whole brain, than is possessed by any other animal.

Weight and Size of the Brain.—The average weight of the brain of man is generally about 50 ounces, or a little over three pounds. The brain of the other sex is generally from four to six ounces lighter. The usual weight is between 46 and 53 ounces in man, and between 41 and 47 ounces in woman. The weight of the brain goes on increasing rapidly up to the seventh year, and more slowly up to the age of forty, when it reaches its maximum weight. With the advance of age and the decline of the mental faculties the weight diminishes. It has been found by examination after death that some persons with brains of light weight have exhibited strong intellectual power, and others who possess heavier brains have not displayed any great claims to intellectual brilliancy. The brain of an idiot seldom weighs more than 23 ounces. Some human brains have been found after death to weigh between 60 and 70 ounces. The quality of the material that makes up a brain must be an element of as much importance as the weight. The brain proper of man is greater in proportion to the weight of the body than that of any other animal. One would think that the elephant or the whale would possess a much greater amount of brain than man, for their heads are so largely developed; but it is found by examination that the brain of the elephant weighs only 120 to 150 or 160 ounces, and of the whale about 80 ounces. This is a much smaller amount of brain, in comparison with the size of the animal, than that of man. The brain of the elephant is only the $\frac{1}{500}$th part of the weight of the whole body, while in man it is about the $\frac{1}{40}$th. The horse's brain is not half that of man's in weight, and yet his body is at least six times as heavy as that of man.

The average proportion of the weight of the brain to that of the body in different classes of animals may be briefly stated, although there are individual exceptions to this statement :

Fishes...:.. 1 to 4000 or 6000.
Reptiles.. 1 to 1500.
Birds .... 1 to 220.
Man.. 1 to 40.
Mammalia, including the higher animals........ 1 to 180.

The development of the brain is in proportion to the intelligence of the animal, so that in the lower scales of creation we find that organ occupying a very small part of the structure of the animal, and parts of the brain that are not concerned in intellectual uses largely developed.

Brains of Different Races.—Several plans have been adopted to determine the actual amount of brain possessed by the different races of mankind. It has been found that the Caucasian race, examples of which are seen in the European and American, has the brain more fully developed than any other. One mode of measuring the capacity of each race was to take the empty skulls of different races of men, weigh them, and, after filling the cavities with small shot, again weigh them, to find which contained the greatest amount of shot. This only showed, however, which would hold the most brain-matter in mass, but did not prove that one possessed more cerebrum or brain proper than the other; and that was the only test of intellect. When we look at the skulls of different races of people we find that the front part of the skull of the Caucasian race admits of a larger space for the brain proper, the seat of intellect.

Facial Angle.—As the portion of the brain immediately in the front part of the skull is the seat of the highest intellectual development, the prominence of the forehead has been considered in man and animals a sign of intelligence. Its measurement has therefore been adopted as a test of the amount of actual brain-power. The facial angle, as it is called, which is formed between the forehead and the face, as shown in Figs. 77, 78, is the test employed. The angle formed by the meeting of the two lines *ab* and *cd* represents the prominence of the front part of the brain, and also the relative amount of intelligence. The great difference in this respect between the fully-developed European and the imperfectly-formed head of the idiot is well exhibited. It will be noticed how defective the front and upper portion of the skull is in the latter, and how little room is left there for occupation by the brain. The angle in one case is almost a right angle; in the other, quite acute. In monkeys the angle is much more acute.

The Brain Proper.—The anatomists have given names to every little point or depression on the surface of the

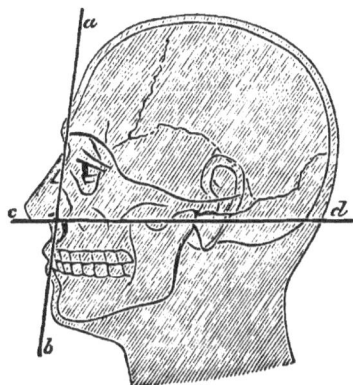
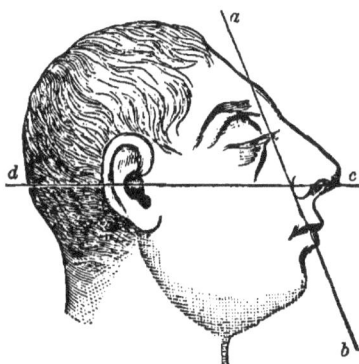

FIG. 77. FIG. 78.

THE FACIAL ANGLE.

A, European; B, idiot.

brain, but with these we have very little to do, as we should be confused with a mere mass of names, without knowing the uses of any of the parts so called. As will be noticed in the illustration (Fig. 76), the cerebrum, or brain proper, is made up of numerous thick, worm-like folds or coils, called *convolu'tions*, and these are divided by several deep cracks or fissures—so deep that in looking at the brain in particular positions it seems as if it were cut in two (Fig. 79, 2, 2). It extends from the forehead in front to the neck behind, and from ear to ear. The chemist has analyzed the brain, but he finds it chiefly made up of fat and phos-

17 *

phorus. The vital principle through which this fatty mass thinks and acts is beyond his power to fathom. The principal fissure of the brain divides it into two parts, called *hem'ispheres* (Fig. 79, 1, 1).

The outer part of each hemisphere is made up of gray or ashy-colored matter, but the interior is almost entirely

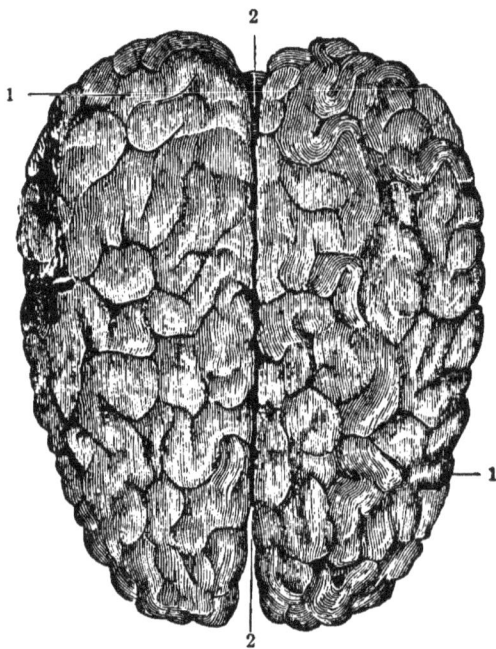

FIG. 79.—EXTERIOR VIEW OF THE BRAIN.

1, 1, right and left hemispheres of brain; 2, 2, fissure, dividing brain into two parts.

white nervous matter. Fibres passing upward from the spinal cord and medulla oblongata connect this with those parts, and fibres pass across also from one side of the cerebrum to the other. At the base of the brain are also other bodies, called *ganglia*, made up of gray matter, through which the fibres pass from below upward to the cerebrum. There is quite a variety of cavities or spaces, with arches and delicate veils and apparently mysterious halls and passageways, which we suppose to be necessary for the perfection of the brain's duties, but we know little or nothing of the uses of each minute part or space, and merely study it as a whole, as the seat of all intellectual power and thought. Judgment and reason and memory have here their seat.

The Lessons of Disease.—Much that is known about the work done by this portion of the brain has been learned from cases which have come into the hands of the surgeon in which the skull has been fractured or the brain injured. In such cases the powers of intellect and memory are destroyed, either for the time being or permanently. A defective memory is one of the first indications of apoplexy or of some other disease or degeneration of the substance of the brain. In idiots it is one of the first symptoms that makes us suspect there is something wrong in the child's brain. Judgment and reason may be deficient for similar reasons, thus proving, from what we know of the effects of disease or injury, what are the functions of the organ when in perfect health. The physician is thus aided by pathology, or a diseased state of an organ, in discovering the nature of its physiology.

Cerebellum.—The little brain, or cerebellum (Figs. 73, 75, 76), is differently arranged from the cerebrum. It is not formed into convolutions or coils, but is, if we may so phrase it, in layers and foliated like the leaves of a book. On making a cut into it for examination a beautiful appearance is presented, called, from its peculiar tree-like arrangement, the *ar'bor vi'tæ*, or "tree of life." This is due to the extension of the white matter on its outer border into the gray matter of its interior, as shown in Fig. 76.

Spinal Cord (Fig. 73).—As already stated, the medulla oblongata is an extension upward of the spinal cord. This cord, or spinal marrow, as it is sometimes called, is really a combination of cords of nervous matter, separated by fissures or cracks running their whole length, and covered by a common sheath of membrane. Before passing upward into the skull the fibres from the right and left sides of the cord intersect one another, and then unite in the medulla oblongata, ascending into the pons Varolii, and thence into the cerebrum, in which they spread out in all directions. It thus seems that the fibres of the right side of the body cross to the left side of the brain, and of the left side to the right side of the brain. Each side of the brain therefore exercises some control over the movements of the opposite side of the body. We have yet much to learn of the physiology and minute structure of this and other parts of the nervous system. Paralysis may result from injury at these points in the brain, and, on account of the crossing of fibres to which we have referred, an injury of the right side of the brain may produce paralysis of the left arm or left leg, or *vice versâ.* There is a similar crossing of some of the fibres of the spinal cord connected with the general sensation of parts, and loss of sensation on the opposite side sometimes occurs.

The Nerves (Figs. 80, 81).—These are continuous threads of nervous tissue which are distributed to all parts of the body. The nerves themselves are made up of collections of nervous filaments bundled together. Each filament is composed of a fine gray thread, surrounded by a soft white substance, enclosed in a thin covering or membrane. On the skin they are very largely distributed. They are generally considered to arise from the brain or the other nervous centres, and convey impressions backward and forward, to and from the surface of the body or from organ to organ within the body. They are the telegraphic wires through which the will and the intellect and life itself make themselves manifest. They seem to be lost in the substance of organs, or, if they terminate on the surface, in minute nervous papil′læ or points. They are either connected in this way with motion, as when distributed to the muscles, or with the nutrition of the body in the interior organs. After a nerve leaves the deeper structures, in which it is protected from injury, and comes to parts in which, by motion of the limbs or by pressure, it might suffer in any way, it becomes covered with a strong white fibrous sheath, which binds together the different filaments. This is called the *nerve-sheath*, or neurilem′ma.

FIG. 80.—A NERVE, NATURAL SIZE.

In the course of some of the nerves are enlargements or knots called *ganglions*. These are especially met with in the great sympathetic nerve, already referred to, which in some mysterious way presides over functions, like secretion, digestion, circulation, etc., that are entirely beyond the control of the will.

The communications between the nerves are not like those which take place between arteries and arteries and veins and veins, by open canals. There is no strict analogy between the bloodvessels

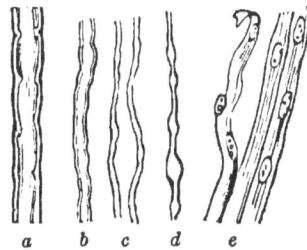

FIG. 81.—NERVE-FIBRES.

a, b, c, d, nerve-tubes of different sizes; *e,* nerve-tube from sympathetic.

and the nerves in this respect, for in the former the fluids actually mix, but when one or two nerves unite to form a single cord there is no mixture of nervous matter, for they merely lie side by side without union. In Fig. 81 are shown the various appearances of the nerve-fibres from different parts. They are very distinct from the nerve-cells previously alluded to.

The Nerves Symmetrical.—All the nerves of the cerebro-spinal axis come off from the base of the brain or from the spinal cord. The nerves come off in pairs—that is, one from each side of the axis, right and left, each corresponding to the other in shape, etc. For illustration: if the body be considered to be divided into two equal parts, right and left, by a section directly through the middle of it, it will be found that the two sides of the body are precisely alike and symmetrical. Every member, every organ, every vessel on one side, with few exceptions, will be found to have a fellow similarly placed on the other side, except, of course, such organs as are single, like the liver and heart, and occupy portions of both sides. The same is true of the nerves of the body, a similar symmetry existing, the nerves issuing in pairs and from each side.

It is usual among anatomists to name the nerves in the order in which they come off from the cerebro-spinal axis, beginning with the brain, but they do not continue calling them by number in regular series all the way down the spinal cord. Those which come directly from the brain and spinal cord number altogether forty-three pairs, but those which issue from the base of the brain are called *cra'nial nerves*, because they come from the cranium or skull, and those from the spinal region, *spinal nerves*. It is usual to say that there are twelve pairs of cranial nerves and thirty-one pairs of spinal nerves. The former, coming from the brain, pass out to various organs through openings in the skull or bony covering of the brain, while the spinal nerves emerge through openings in the bony spinal column. Some physiologists take an opposite view, and consider that the nerves originate in the skin and muscles in different parts of the body, and pass inward to the spinal cord, instead of originating in the nervous centres and passing outward to the skin, etc. Under this view they describe the spinal cord as made up of filaments derived from all the nerves of the external parts and the limbs. The brain itself they consider a round nervous mass or expansion formed by the upper part of the spinal cord projected into the skull.

Cranial Nerves.—It is not necessary to go into a description of these cranial nerves except in a general way. They supply the organs of sight and smell and hearing, the face (Fig. 82), the mouth and tongue and throat; and some of them are distributed to the heart and lungs and stomach. The first pair is called the olfactory nerve (from two Latin words which mean "causing smell,") because it is the nerve of smell, sent to the nose to appreciate odors. The second nerve is the optic nerve, because it is the nerve of vision, which if injured destroys sight, no matter how perfect the rest of the eye may be.

The *fifth pair of nerves* is distributed to the face, the eye, the nose, the mouth, etc., and gives feeling and general sensibility to the parts it supplies, and adds also to the perfection of the senses of taste, touch, etc. It is one of the most important pairs of nerves in the body, and many of its branches are represented in Fig. 82. A recent writer in an English magazine states as a curious fact that people of all nations are accustomed, when in any difficulty, to stimulate one or another branch of the fifth nerve and quicken their mental processes. Thus, some persons when puzzled scratch their heads, others rub their foreheads, and others stroke or

pull their beards, thus stimulating branches of the same
nerves. Many when thinking have a habit of striking their
fingers against their noses, and thus stimulating the branches
distributed to the skin, while some people stimulate the

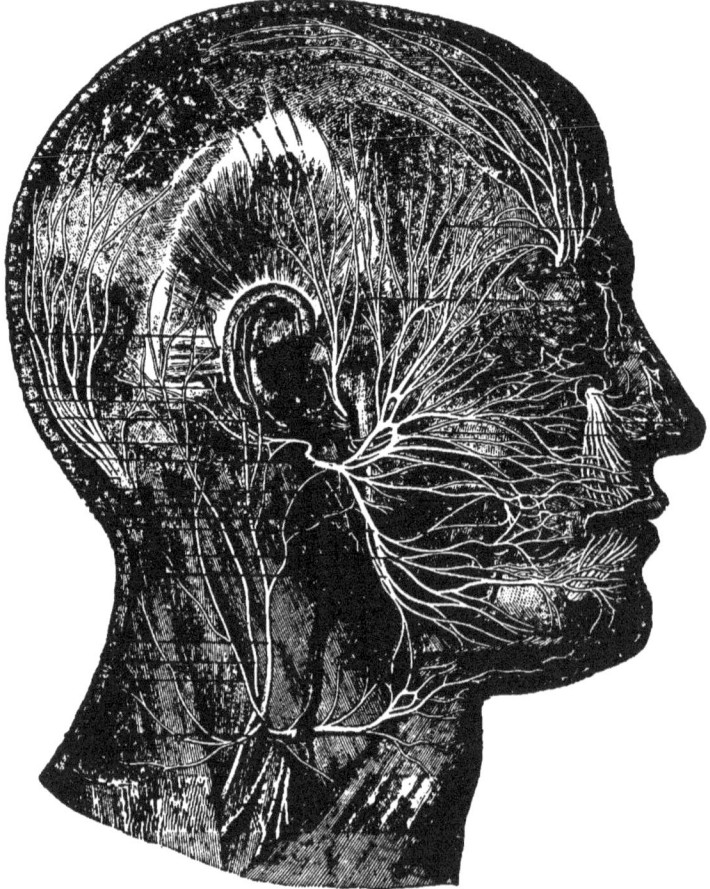

FIG. 82.—SUPERFICIAL NERVES OF THE FACE AND NECK.
(The guiding lines indicate the most important nerves and their branches. Their
names are purposely omitted, as unimportant to the reader.)

branches distributed to the mucous membrane of the nose
by taking snuff. One man will eat figs while composing a
leading article, others will smoke cigarettes, and others sip

brandy and water. By these means they stimulate the branches of the fifth nerve distributed to the tongue and mouth, and thus reflexly excite their brains. Neuralgia of the face is dependent on some irritation or disease of this fifth pair of nerves. An examination of the structure of these nerves gives no insight into the duties they have to perform.

Some of the cranial nerves are connected with motion of the face, as in the expression of emotion, etc. ; some are merely nerves of special senses, like hearing and sight, and have no other function. These would give no indication of pain if cut or injured. When several nerves are distributed to the same organ—the tongue, for example—the duty of one of them will be to supply the special sense, as taste, for example ; of another, to produce motion through the muscles of that organ ; of a third, to give general sensibility to it, as evidenced by pain when it is scraped or injured in any way.

Spinal Nerves.—These, as they issue from each side of the spinal cord, have two roots, which, approaching each other, join to form the nerve before its exit through the opening in the spinal column previously spoken of. It may be here stated that the substance of the spinal cord consists of gray and white matter, but differently arranged from that observed in the brain. Here the outer or cortical part of the nervous matter is the white substance, while the interior is the gray. The amount of white matter is greatly in excess of the gray. Three coats cover the cord, similar to those of the brain, and the cerebro-spinal fluid fills a similar space. The fibres which make up the larger or posterior root of each spinal nerve are nerves of sensation, while those which form the anterior or lesser root are nerves of motion. On the posterior root is a small

18

ganglion or swelling (Fig. 83). There are two bundles or

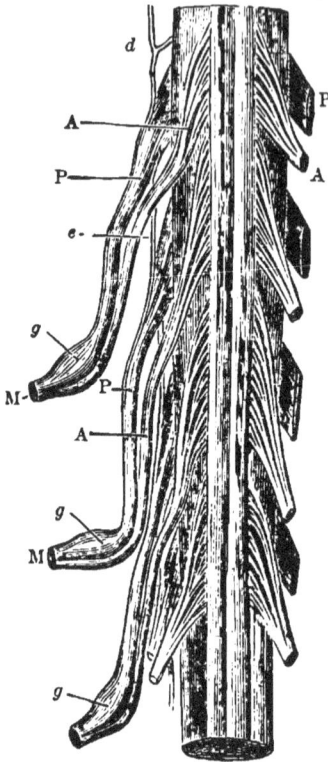

columns of white matter on the back of the spinal cord, with which the sensitive fibres of the posterior roots are connected, and two columns in front, with which the motor fibres of the anterior roots are connected. These are known as the *posterior and anterior columns of the cord*.

Effect of Injury on Nerves.— When the spinal cord is injured or destroyed, or when the spinal column is broken, motion and sensation may, either one or both, be affected by the wound or by pressure of the broken bones. Every one is familiar with the sensation of a hand or a foot being "asleep." This is caused by temporary pressure on the nerves going to those parts from sitting or leaning upon them, numbness and want of power being the chief facts noticed. When a nerve is cut or torn, motion and sensation are both lost to the parts be-

FIG. 83.—ORIGINS OF THE SPINAL NERVES.

The spinal cord is represented in the cut. A, A, A, anterior roots arising by minute divisions which unite afterward to form the fibres of the root; P, P, P, posterior roots; c, d, filaments passing between the posterior roots; g, g, g, ganglions of posterior roots; M, M, mixed nerves formed by reunion of two roots. (The size of the roots is exaggerated.)

yond. The muscles cannot act, for the telegraphic wires are broken and they do not receive the message, and the skin feels dead to all sensation because the sensory nerves do not go beyond the gap thus made. Paralysis, or loss

of power of motion and of sensation, results, but if a portion of the nerves escapes injury the parts supplied by them
will not be paralyzed. Motion and sensation are both lost
at the same time, because both kinds of fibres lie side by
side in the nerves affected. The kind and degree of the
paralysis vary with the part of the spine involved. Thus,
if the nerves in the middle of the back are injured, paralysis
of the legs only will result. The arms would only become
paralyzed if the upper parts of the spine, from which the
nerves supplying the arm are given off, are involved. Motion and sensibility are generally lost together, but if the
anterior column of the cord is alone affected, there will
only be loss of motion, and of sensation in the case of the
posterior column.

When a nerve is cut across by accident or by the knife
of the surgeon in operating, its power is not necessarily
destroyed for ever. The nerves are so liberally supplied
to all parts of the body that the slightest injury must
affect some of them. Permanent loss of motion or of sensation very rarely results, because the nerves heal up by
growing together again, not exactly as the skin or the flesh
does, and not so rapidly.

How the Spinal Cord Terminates.—The spinal cord does
not descend in a continuous mass the whole length of the
spinal column. At a certain distance down it ends in a
bundle of nerves which pass off after the manner of a
horse's tail, and is therefore called the *caud'a equi'na*—
which is the Latin for a horse's tail—the nerves from
which continue to pass out from the holes in the spine in
the same way as the nerves in the upper part of the cord.
The nerves which are distributed from the spinal marrow
convey sensibility and power of motion to the most distant points. Where they terminate they are sometimes so

minute as to defy the powers of the microscope. In their course they are a collection of nerve-fibres contained in a common sheath or neurilemma, as it is called.

It seems impossible to find out, even with the microscope, whether each nervous thread goes on continuously from the brain through the spinal cord, or whether these parts and the nerves are all connected with one another by other fibres.

Nerves of Respiration.—Some of the spinal nerves are distributed to the muscles of respiration, such as the intercos'tal muscles—those between the ribs—and to the diaphragm. We have already seen how important this muscle is in the act of breathing—ordinary gentle breathing being effected by its rise and fall, without calling into full play the other muscles of the chest and abdomen. The intercostal muscles, by their action, raise and lower the ribs. We can readily imagine the result of an injury to the spine which would paralyze these muscles. If both nerves are cut, or the spinal cord is divided above them, death must result from stoppage of respiration, for the diaphragm will no longer rise and fall, being deprived of its nerve of motion, and the intercostal muscles will not aid the ribs to move.

The Great Sympathetic Nerve (Fig. 84).—It has already been stated that a series of small bodies or ganglia of nervous substance, connected together by nervous cords and threads, and communicating with the other great nervous systems, is found spreading itself everywhere, through the chest and abdomen particularly. As will be seen in Fig. 84, it is principally distributed in a symmetrical manner on each side of the middle line in front of the spinal column. It has a seat also in the brain, from which muscles of the eye and ear and other organs of sense are supplied.

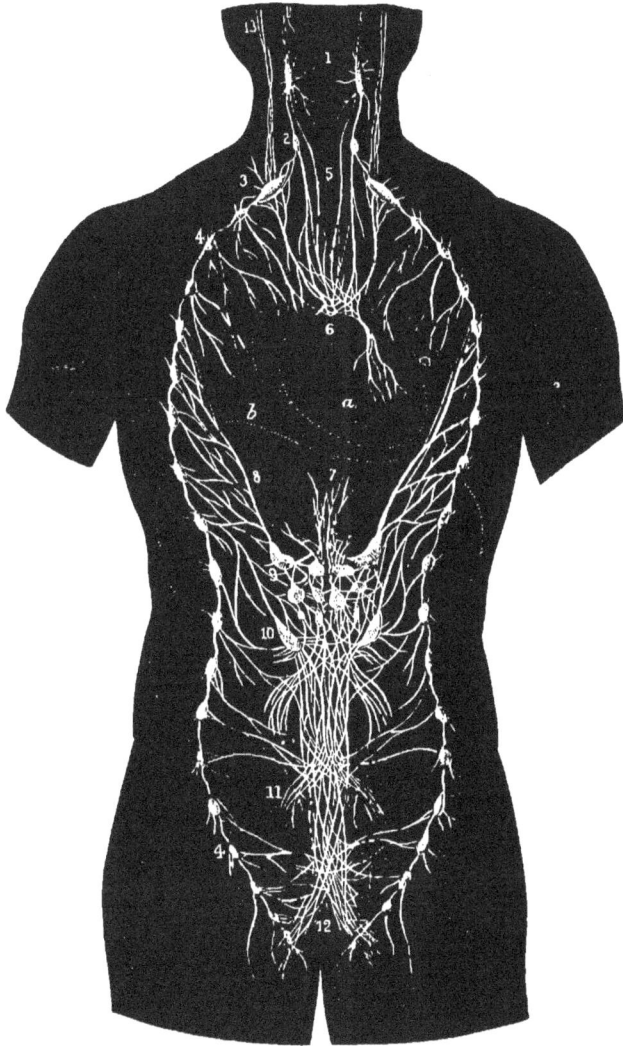

FIG. 84.—THE GREAT SYMPATHETIC NERVE.

1, 2, 3, ganglia in the neck; 4, spinal ganglia; 5, branches in neck and chest going to heart; 6, nerves to heart; 7, nerves about diaphragm; 8, nerve to digestive organs; 9, semilunar ganglion; 10, 11, 12, masses of nerves to abdomen; 13, small nerves going with arteries to brain. Dotted lines indicate the position of a, the heart, and b, the diaphragm.

18 *   O

It presides over involuntary functions associated with the maintenance of life, such as digestion, respiration, circulation, secretion, etc., but has nothing to do with the great mental or moral acts or with the voluntary motion of parts. These are influenced by the cerebro-spinal system.

The small bodies, or ganglia, which are found in the course of the sympathetic nerves (Fig. 84) have been generally considered by physiologists as centres of nerve-force—little brains, if we may so call them—which originate nervous power and send it speeding along the nervous wires to the various organs. It seems a confirmation of its influence in carrying on such important labors as are involved in digestion, respiration, etc. that the sympathetic system of nerves is developed before the other portions of the nervous system. It is a wise provision that places these important processes beyond the power of the will; otherwise our lives would be made miserable with efforts of constant attention directed to the execution of these vital acts. Sympathetic branches are supplied to the muscular coats of the bloodvessels, which have the power of regulating the calibre or capacity of such vessels, and therefore the quantity of blood they may contain. A familiar example of this influence is seen in the act of blushing, or in the pallor which sometimes instantaneously covers the face. In some of the lower forms of animal life it has a separate and distinct existence in the entire absence of brain or spinal cord.

Functions of the Nervous System.—The greatly diversified arrangement of the brain and spinal cord shows that the nervous system is intended to fulfil a varied series of duties in carrying on life. Otherwise, what would be the necessity of this complicated structure, this division of two kinds of matter, gray and white, or the variety in form and material of the different parts composing the nervous system? Physiology cannot here trace the exact relation of parts to the duties to be performed, as can be so readily done in machinery operated by physical causes. It has discovered much, but it is restricted in its inquiries. Living and dead animals have been experimented upon, with the view of increasing our knowledge, and diseases and injuries of the nervous system of man have also given us information as to the functions of the nervous system. The effort has often been made to find out the uses of every prominent point in the brain—for example, by stimulating it with electricity, etc.—but without much result. What has been learned from all these sources may be briefly stated.

Functions of the Nerves.—When an impression is made on any part from without—as by a blow, or by temperature, as heat or cold, etc.—it is received by the minute nervous threads which are present in every part, and conveyed along them to the brain, which is the great central organ of sensation and perception. The impressions thus conveyed are the *sensations* with which we are so familiar. The brain is also the seat of volition, or of the will; and the nerves, returning to the part affected, have the power of receiving another impression from the brain, and of transmitting it to a particular muscle or organ. They can thus call into motion, under the influence of the will, the muscles which it selects as the agents of its exercise, or stimulate other organs to action. It will thus be seen that there are two sets of nerves—already described—endowed with different functions, one set called *sens'ory nerves*, or nerves of sensation; the other set communicating motion, and called *motor nerves*. The reason of this difference of function cannot be explained. They may run closely to-

gether in the same nervous sheath, but without any communication with one another.

If the nerve connecting the limb of an animal with the
nervous centres, as the brain or spinal cord, be exposed
and mechanically irritated, two effects will follow: pain
will be manifested, and the limb will be thrown into
spasms. So, too, if the hand be accidentally brought into
contact with a hot substance, a sensation of pain is communicated to the brain by the nerves of sensation, and the
hand is at once removed by an impulse sent through the
nerves of motion. The fibres of motor nerves are distributed through the substance of the muscles. When we
hereafter describe the Skin we will understand more clearly
the mode in which the nerves of sensation arise on the surface of the body before passing inward from the skin to
the nervous centres.

Reflex Actions of the Spinal Cord.—These are actions
continually taking place in the body from impressions
made from without, which are not under the control of
the judgment or the will. They show that when the brain
is not active, as during sleep, or after injuries which cut
off communication between the brain and the external part,
the spinal cord can take notice of the impression, and act
in response to it. Sometimes voluntary motion may be
destroyed in some way, and yet, if any irritation be applied
to the skin of the part which has been paralyzed, the
muscles will act temporarily independent of the will. It
has been noticed in those who have entirely lost the use
of their lower extremities that a spasmodic action of the
muscles of the affected part may take place from the
impression of cold or the striking of some other body
against it. The brain does not control the action, for its
communication has been cut off. The power to regulate it

lies in the gray matter of the spinal cord, which is deeply seated and runs the whole length of the cord.

To keep up this communication requires that the sensory and motor fibres, already described in their connection with the columns of the spinal cord and the brain, should be connected also with the gray matter of the cord; and this we find to be the case. Movements made in response to this action of the spinal cord are said to be *reflex*, because the impression is made on the surface of the skin, and thence conveyed to the cord, which reflects or sends back the motor impulse to the muscles. The action is called *reflex action*. The nature of this movement is illustrated in the accompanying diagram (Fig. 85).

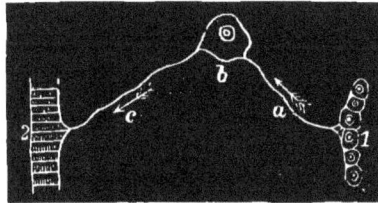

FIG. 85.—SIMPLE REFLEX ACTION.
1, sensory surface; 2, muscle; *a*, sensory nerve; *b*, nerve-cell; *c*, motor nerve. (The nervous influence travels in the direction indicated by the arrows.)

The importance of having such a power assigned to the spinal cord, independent of the will or of sensation, is seen in every-day life. The rapid and involuntary manner in which we throw up the arm as a shield from immediate danger, as a fall or a blow, is an illustration in point. If we seize a substance that is too hot to hold, we drop it involuntarily through this very reflex action. During sleep the mouth will often receive and the throat swallow water involuntarily when placed to the lips. The body also turns in bed during sleep, not at all under the control of the will, but as a sequence of this reflex action alluded to. What are commonly known as spasms or fits are examples of an intensity of reflex action of the spinal cord. The muscles are sometimes, as in lockjaw and hydrophobia, called into violent and uncontrollable exercise, and the action of the muscles is painfully and intensely increased under the stimulus sent out to them from the spinal cord.

From all these facts we learn that the spinal cord not only acts as a messenger between the external parts of the body and the brain, but that it has also an independent duty of its own, and effects muscular movements that are entirely beyond the control of the will.

Functions of the Medulla Oblongata.—We have already shown that this portion of the nervous system is the connecting link between the brain and the spinal cord. It contains a larger amount of gray matter than we find in the spinal cord. Its duties are chiefly to preside over the functions of respiration and deglutition. It is a conductor of nervous impressions, as the cord also is. Motor impressions—those which result in the movement of muscles—pass along its anterior columns. All sensory impressions—those which call up sensation in the brain—pass along its posterior columns.

The fact that respiration and deglutition or swallowing are controlled by the medulla oblongata shows how absolutely necessary this nervous centre is to the maintenance of life. Experiments on animals have shown that almost all the brain may be gradually removed without destroying respiration or life. The spinal cord can also be cut away in animals from below upward until near the throat the phrenic nerve is reached, and this is the nerve distributed to the muscle called the diaphragm, whose rise and fall in respiration have already been alluded to (p. 130); but the moment this nerve is cut or the medulla oblongata is impaired death ensues. Paralysis would of course result in similar experiments on the brain or spinal cord, but not necessarily death.

This portion of the nervous system is also capable of the

most important reflex actions connected with respiration, as is seen where, from any cause, such as lockjaw, etc., the muscles connected with breathing become fixed and immovable and respiration is impossible.

Functions of the Cerebellum.—The duty performed by this part of the brain was for a long time a puzzle to physiologists. It could only be determined by experiments on living animals, as may indeed be remarked of other parts of the nervous system. It seems to be clearly established that it is the nervous centre which regulates and keeps in order the motions of man and animals. In the experiments alluded to the sensation remained, but the animal could not move, fly, or walk, or even stand. By the action of the cerebellum the movements of those muscles which are under the control of the will are carried on harmoniously. The more developed the quality or variety of muscular action, the more developed is this organ found to be. This fact explains the high state of development it attains in man, whose muscular system is attended with such varied and complicated movements.

The cerebellum may be briefly stated, then, to be a regulator of muscular movements. In some of the experiments on pigeons—which seemed essential at first to establish an important fact, but which can hardly be considered necessary now merely to gratify morbid curiosity—it was found that if the cerebrum or brain proper was taken away from one of these birds the animal would remain firm on its feet, while a pigeon from which the cerebellum was only partially removed would exhibit the unsteady and uncertain gait of drunkenness. Sensibility is not lost in the latter case. The cerebellum does not originate movements, but it regulates and gives precision to them. This is another case in which disease or injury aids

us in our knowledge of physiology—a point on which we
have insisted from the very outset of this work (p. 11).
When the cerebellum is affected in man—and this seldom
occurs—there is always unsteady movement and a tendency
to backward or other unusual motion. It is believed also
that the regular movement of the eyeballs is controlled by
the cerebellum.

Functions of the Cerebrum.—As previously stated, the
cerebrum or brain proper is the great organ of thought,
sensation, and intelligence. The exercise of judgment and
reasoning power belongs to it alone. The brain of man is
in size and development far superior to that of all other
animals. When we come to speak of the nervous system
of other animals than man we shall understand more clearly
wherein this distinction lies. It may here be stated, in an-
ticipation, that as a general rule an accurate estimate may
be formed of the intelligence of animals by comparing the
size of the cerebrum or brain proper with that of the spinal
cord and the ganglia or nervous masses at its upper portion.
Although reflex movements, such as those already alluded
to, may be carried on for some time even in animals that
are born without a brain, and sufficient motion take place
to sustain life, all power of voluntary motion is absent.
How much of the brain, whether all of it or only a part,
is necessary to effect this exercise of voluntary motion, it
is difficult to say, for small amounts of the substance of the
brain have been lost by accident, and yet the power has
not been lessened in any way. It is not many years since
an iron bar used in blasting near a New England village
passed directly through the brain of a workman without
destroying life or very materially interfering with his movements or with the exercise of his intellect or will.

Physiology has not yet determined what are the uses of

each separate lobe of the brain as distinct from its fellows. It has been suggested that it was originally divided into portions, rather than left to remain a single tremulous mass, in order to provide against the possibility of injury or disease of one lobe involving the whole brain-mass in general destruction, and thus endangering life. It has been thought, at any rate, that the two hemispheres may act—that is, think —separately as well as together. Certainly there are moments in the experience of every one when the mind, seemingly intent on a single theme, as in reading, will wander off to a hundred other fancies. Perhaps in time human ingenuity will be able to trace the two different lines of thought in which the mind then indulges to the separate hemispheres or the individual lobes as their starting-point.

Let it be distinctly understood, in regard to the movements of the body, that the cerebrum originates and controls the voluntary motions or movements of the body, while the cerebellum is the organ which harmonizes them, and that reflex movements, such as those previously alluded to, which are excited entirely independent of the will, are connected with the spinal cord and the sympathetic system of nerves.

It has been urged against the accuracy of our knowledge of physiology derived from experiments on animals, and by analogy applied to man, that independent of its cruelty and severity the loss of blood may act as a severe shock to the nervous system. A new plan has recently been practised of developing responses from the brain by the action of electricity applied to its various parts. It was found that if one particular spot on the brain was stimulated the muscles of the neck were called into action; if another spot, the muscles of the face or of the eyeballs or of the leg responded. Nothing is yet certain here, however; we may

19

hope to learn much in the future. There is a form of disease of the brain in which the patient finds it not only impossible to express his thoughts in words, but sometimes may employ the wrong word to express his thoughts, as when he wishes a knife and asks for a spoon. In such a case one of the convolutions of the left side of the brain is usually found to be affected.

Phrenology.—As already stated, it has been thought by many that the brain may be made up of a collection of different organs, each of which may be endowed with special powers or qualities. This is called *phrenol'ogy* (from two Greek words meaning "doctrine or description of the mind"); but the word is very often understood to imply the examination of the elevations and depressions on the skull, with the view of learning from them something of the peculiarities of mind of the individual. The attempt has been made to map out the surface of the skull with tracings upon it of mental characteristics, under the idea that there is some relation between them and the formation of the cranial bones. While it is very possible that in time we may be able to assign functions to each portion of the brain, there seems little ground for believing in the skull as an index of its contents.

Sleep.—It is a strange fact that the physiology of sleep is so little understood at this late day. We know that the active exercise of the brain is temporarily suspended at that time, and that respiration, digestion, secretion, and circulation go on as during the waking state, although not with the same activity; that the object of sleep is rest, that it is likely to come on regularly at fixed hours, and that it can sometimes be controlled by the will, and is often a matter of habit. The amount of sleep necessary to man depends on the age, habit, and condition of the individual. The very young and the very old require the most sleep—the younger, because the vital processes are in such a state of activity that rest is an absolute need; the older, because they are so feeble that sleep and repose are indispensable. During sleep the pulse becomes slower, the number of respirations fewer, the breathing deeper, digestion less active, and a smaller amount of carbonic acid is given off from the lungs. The effect on the brain is to deaden the perceptions, to produce confusion of ideas and a loss of mental balance. So far as relates to the exact cause of sleep we are yet in the dark, but it seems to be the fact that there is less blood in the vessels of the brain at that time. Eight hours of sleep would seem to be the amount which is usually required by the adult, but this will of course be governed by habit and other circumstances.

When the brain becomes at all active during sleep *dreaming* results, but such activity never occurs during profound slumber. It may be due to disturbances of the digestive organs, as after a full meal, and when exaggerated by great oppression from this cause may give rise to *nightmare*, as it has been called.

Summary.—We may briefly sum up the facts known to us in relation to the nervous system. The nervous elements with which we have to deal are the two divisions of the nerve-substance into nerve-cells and nerve-tubes or fibres. The nerve-tubes serve as conductors of the nervous fluid, being likened in this respect to the action of electricity, although differing greatly from it. The nervous matter is of two kinds, gray and white; the gray is found in the outer part of the cerebrum or brain proper and in the inner part of the spinal cord, and especially in the nervous masses, large and small, called ganglia, of which the gray matter forms the greater part. The white substance is found in

the interior of the brain, the outer portion of the spinal
cord, and in the nerves. It may be stated as a rule that the
gray matter originates impressions or actions, and the white
matter conducts or conveys them away.

The general nervous system is divided into a cerebro-
spinal system and a sympathetic system. The cerebro-
spinal system includes the cerebrum, cerebellum, medulla
oblongata, spinal cord, and the nerves associated with them.
The great sympathetic is a chain of ganglia beginning at
the brain and running the whole length of the chest and
abdomen, with nerves going off to the various organs, to
the vessels, and to the cerebro-spinal system. The cerebrum, or brain proper, is the seat of sensation, judgment,
and intellect, and controls voluntary motion. The gray
matter is the originator of mental power, and the higher in
the scale the animal is, the greater the amount of this substance. The cerebellum harmonizes the movements of the
body. The great sympathetic nerve, by its distribution to
the heart, the various glands and organs, brings the various
parts of the body in sympathy with one another, and by
its distribution to the bloodvessels changes their capacity
and regulates the amount of blood sent to them.

The medulla oblongata is the centre of the respiratory
movements, and most of the cranial nerves, except those
of sight and smell, take their origin from it or from the
portions of the brain close to it. The spinal nerves have
two roots—one anterior, the other posterior,—the posterior
root containing only sensory fibres, or fibres connected with
sensation, the anterior root only motor fibres, or fibres connected with motion. Some of the cranial nerves—those
derived from the base of the brain—are concerned in sensation—either general sensation, such as would give rise
to pain if injured, or special sensation, as indicated by the

senses of hearing, sight, etc.—or in motion. As the structure of a nerve reveals to us no difference to account for this variation of function, we must suppose that the gray matter of the brain with which the nerves communicate must have something to do with this peculiarity.

The gray matter in the centre of the spinal cord has the power of originating reflex action, and of creating reflex movements, without calling upon the brain. General impressions are transmitted to the brain along the spinal cord and medulla oblongata, and the brain acts in response. The spinal cord, in its reflex action, regulates involuntary movements connected with the nutrition of the body entirely independent of the brain, and also involuntary movements of voluntary muscles, as when we throw up the arm to shield us from danger. It will be readily seen how much labor is thus saved the brain proper, which is left to the performance of its high duties and intellectual work, in many respects untrammelled by attention to minor details of every-day life.

There is very little known of the functions of the nervous masses, called cranial ganglia, at the base of the brain, except that they resemble the spinal cord in the fact that they are concerned in sensation and motion.

The Nervous System of Animals.—Without some knowledge of the arrangement of the brain, spinal cord, and nerves in animals lower in the scale than man, physiology would have been unable to reveal so much in regard to his nervous structure. When we look through the whole range of animal life, we find a general plan running through all, differing in details according to the necessities of each animal. It is usual to compare man with the mammalia generally—that is, animals next him in the scale—and with birds, fishes, and reptiles. All these are

19 *

constructed with the same leading parts as already described in man. They have a cerebrum or brain proper,
cerebellum, medulla oblongata, and spinal cord, with their
proper nerves, and they have a sympathetic or ganglionic

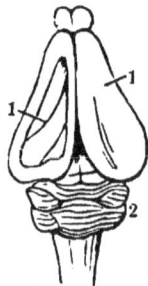

FIG. 86.—BRAIN OF A RABBIT.

1, hemisphere of brain; 2, cerebellum.
The middle lobe of the cerebellum is
quite large.

FIG. 87.—BRAIN OF A CAT.

The convolutions are very distinct, and
the middle lobe of the cerebellum is
small.

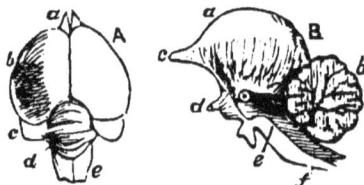

FIG. 88.—BRAIN OF A PIGEON.

A, view from above; B, lateral view of the brain, cut in half.

A. *a*, olfactory lobes; *b*, lobes of brain; *c*, optic lobes; *d*, cerebellum· *e*, medulla
oblongata.

B. *a*, cerebrum; *b*, cerebellum *c*, olfactory nerves; *d*, optic nerves; *e*, medulla;
f, spinal cord.

system, all of which possess the properties ascribed to them
in the human apparatus.

The first and very important point of difference is in
the *hemispheres* of the brain. In man these attain their
highest development. In him they are not only more ex-
tensive, but the convolutions are deeper and so well-marked
and numerous as to admit of a much larger amount of
room for brain-matter. They cover completely the little

brain, or cerebellum, and all other parts of the brain-mass near them ; but as we descend in the scale of animals we find them covering them less and less, until the cerebellum is seen to project decidedly beyond the cerebrum. This will be noticed in the accompanying illustrations (Figs. 86, 87, 88). Other portions of the interior of the brain are less perfectly developed, and some parts are entirely absent.

In the very lowest forms of animal life, such as sponges and the diminutive creatures in fluids which have been called infuso'ria, there is total absence of a nervous system, so far as we can determine. In the sea-anemone a trace of a nervous system is seen as a little knot of nervous matter, like a ganglion, with fibres radiating from it like nerves.

Some insects and shell-fish, which have no spinal column, differ of course in the interior arrangement of the nervous system, ganglia or nervous centres taking the place of a spinal nervous apparatus, and acting as brains (Fig. 89). In some of those animals which have no spinal column—the star-fish, for example—there is a ring of nervous substance around the œsophagus which sends off branches to all the rays, and this is its whole nervous system. In the mollusca—animals of which the oyster is an example—there are usually

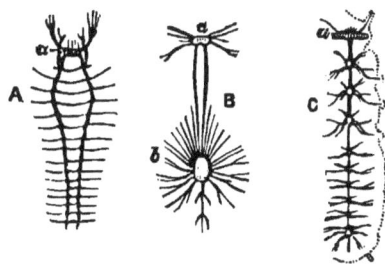

FIG. 89.—TYPICAL FORMS OF NERVOUS SYSTEM IN INVERTEBRATES.

A, nervous system of a serpula, a marine annelide; a, cephalic ganglion. B, nervous system of a crab; a, cephalic ganglion; b, mass of ventral ganglia fused together. C, nervous system of an ant; a, cephalic ganglion.

three ganglions, in the head, foot, and above the alimentary canal—with nerves radiating from them, in this respect resembling the sympathetic or ganglionic system of man. They have sensation and motion and internal organs, all in simple degree, but sufficient to show a similar design running through the animal organization. In *insects*, which as we know vary so greatly among themselves in size and structure, the nervous system is not a constant arrangement. In all of them there is a distribution of nerves to the organs of digestion and circulation, similar to the sympathetic system in man. In some of

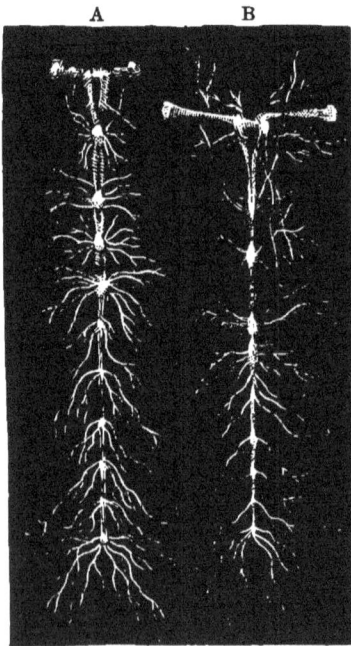

FIG. 90.—NERVOUS SYSTEM OF INSECTS.

A, grasshopper; B, stag-beetle.

these little animals, which are made up of a series of rings placed one above the other, the nervous ganglions are arranged like a chain, from which nerves are given off. Some idea of the nervous system of these animals may be formed from the accompanying illustration (Fig. 90). It is rather a singular fact that all those animals which are devoid of a spinal column—invert'ebrate animals—have their nervous system along the under or ventral part of the body, as it is called (from *venter*, "the stomach").

In vert'ebrate animals—that is, animals which have a spinal column—the nervous system is along the back part of the body. They have, in other words, a cerebro-spinal axis, such as we have described in man, consisting of brain and spinal cord. There is not much difference between one of these animals and another in the arrangement of the spinal cord, the chief distinction being in the development of the brain. The hemispheres of the brain, for instance, are very small in some fishes, and much longer and broader in the tortoise. In birds (Fig. 88) the hemispheres are still larger, and the middle portion of the cerebellum is all that we see of that organ. In some animals—the rabbit (Fig. 86), for instance—the convolutions are almost entirely absent. As we ascend in the scale (Fig. 87), the hemispheres become still larger, the convolutions begin to make their appearance, and gradually become more numerous and deeper. In the dog, this increase of number and depth is very apparent. Of the elephant, this is still more emphatically true.

It is hardly complimentary to man, but it is nevertheless the fact, that in the chimpanzee and orang-outang there is noticeable a decided approach to the structure and shape of the human brain in the development of the hemispheres, proportionate size of the convolutions, etc. The brain of these animals is about one-fourth that of man in size.

Such is briefly a statement of the chief facts that it seems necessary to impart of the knowledge thus far acquired of the structure and functions of the nervous system. The subject is intricate at best, and the most cultivated minds in and out of the medical profession are still puzzled over many undiscovered points of its physiology. Memory and language, for instance, have not yet been assigned their exact place, although the faculty of language is said by some to have its tenement in a small portion of the brain near the left temple. If this should prove to be so, we may hope in time to give to other qualities, accomplishments, and virtues "a local habitation and a name" in tracts of brain-substance long since explored, but hitherto deemed barren and desolate. The reader must

P

not neglect the study of the nervous system from any thought of its difficulties, but, like those who have preceded him for many, many years in the search after the unknown, endeavor to master them by patience and study. He will thus acquire the amount of outline knowledge with which every well-educated student should become familiar.

QUESTIONS.

Is a nervous system possessed by the vegetable?
What functions belong to both plants and animals?
What is the object of a nervous system?
Is the nervous system under control of the will?
What is the cerebro-spinal system?
What are the two great nervous centres controlling it?
What are the nerves?
How do we know that they are largely distributed through the body?
What is the sympathetic system?
What is the ganglionic system?
What are its uses?
What are ganglia? How are they connected together?
Do the sympathetic nerves communicate with other nerves? Why?
How is the gastric juice excited by sympathy?
What two kinds of nervous matter are there?
Which is usually found in the nerves?
What is the appearance of the gray matter?
Do we have the white or the gray matter in the ganglia?
In the brain how are the white and gray matters arranged?
Which kind of matter—the gray or white—originates nervous power?
Which conveys it?
What is the cerebro-spinal fluid?
What are its uses?
Into what three parts is the brain divisible?
What is meant by the words medulla oblongata?
What is the pons Varolii, and why is it so called?
How much of the brain is the brain proper or cerebrum?
What part of the brain is the seat of intellect?

What is the average weight of the brain in man and woman?

How does the weight vary with age?

What is said of the weight of an idiot's brain?

How does the weight of the human brain compare with that of the elephant and whale?

What proportion does the weight of the brain bear to the weight of the body in man? In the elephant? In the horse? In fishes? In birds?

What races have the greatest development of brain?

How is the capacity of the skull ascertained?

What is the facial angle? What information does it convey?

What are the convolutions of the brain?

What is the chemical composition of the brain?

What are the hemispheres of the brain?

How are the gray and the white matters arranged in the brain?

How are the brain and spinal cord connected?

What may we learn of the uses of the brain from accident or disease?

What is the cerebellum? What is its general arrangement?

What is the arbor vitæ of the cerebellum?

What is the general arrangement of the spinal cord?

What is the effect of the crossing of fibres from the cord?

What are the nerves?

Of what materials are the nerves formed?

What are the uses of the nerves?

How are the nerves protected from injury?

What are ganglions?

What portion of the nervous system is made up of ganglions?

What processes or functions does the sympathetic system control?

Do the nerves communicate with each other in the same way that bloodvessels do?

What is the cerebro-spinal axis?

What portion is included in the skull? In the spine?

How are these parts protected from injury?

With how many coats or membranes is the brain covered? The spinal cord?

Where do the cerebro-spinal nerves come off?

Are they single?

How is the body divided symmetrically?

How many spinal nerves are there? How many cranial nerves?

What is believed by some physiologists to be a different origin and course of the nerves?

How do the cranial nerves pass out of the skull? How do the spinal nerves pass out?

What organs are supplied by the cranial nerves?

What is the nerve of smell? Of vision?

What organs are supplied by the fifth pair of nerves?

What are the uses of this nerve?

What are the duties of the other cranial nerves?

When several nerves are distributed to an organ—the tongue, for example—what different duties have they?

What are the roots of spinal nerves?

How are the gray and white matters arranged in the spinal cord? Which is in excess?

Is the posterior root connected with sensation or with motion? The anterior root?

What nervous body is there on the posterior root?

What are the columns of the spinal cord?

What effect has injury of the cord on sensation or motion?

What is meant by a part, as the foot, being asleep?

Why is motion of the muscles interfered with when a nerve is cut?

Why does the skin feel numb?

What is paralysis?

If the posterior column of the cord is injured, is motion or sensation interfered with?

How does the lower part of the spinal cord terminate?

What muscles connected with respiration are supplied with spinal nerves?

What is the muscle chiefly concerned in gentle breathing?

How would respiration be affected by injury of these nerves?

In what portion of the body is the greater part of the sympathetic nerve placed?

What functions or processes are under control of the sympathetic nerve?

What is the duty of the ganglia connected with this nerve?

What action of this nerve is blushing an example of?

Does the sympathetic system ever exist alone?

When an impression is made on any part of the body, how is that impression conveyed to and from the brain?

What are the sensations?

What is the great central organ of sensation? Of volition, or the will?

What are the duties of the two kinds of nerves?

What effects on sensation or motion follow if the nerves going to the brain or spinal cord are exposed?

To what organs are motor nerves distributed?

What activity does the spinal cord exhibit during sleep or when the brain is not active?

What part of the spinal cord then acts?

What are reflex movements?

What familiar examples of reflex action are cited?

What, then, are the different duties of the spinal cord?

What is the relation of the medulla oblongata to the brain and spinal cord?

What functions or processes does the medulla oblongata preside over?

Through what columns of this medulla do motor and sensory impressions travel?

Which is the most necessary to the life of an animal, the brain, spinal cord, or medulla oblongata?

What is the function or duty of the cerebellum?

What has been the result of experiments on the cerebellum of animals?

Has the size of the cerebellum any relation to the muscular power?

Does the cerebellum originate motion?

What portion of the eye is regulated by the cerebellum?

What are the functions or duties of the cerebrum?

How does the brain of man compare with that of other animals?

What influence has the brain over motion?

Does injury to the brain destroy its power?

What are the functions of the separate lobes of the brain?

Do the two hemispheres act together or separately?

What portion of the nervous system originates and controls the movements of the body? What portion harmonizes them? What portion presides over the reflex movements?

What objection may be made to the accuracy of experiments on animals?

What is the effect of electricity applied to different points on the brain?

What is phrenology?

What is it usually understood to include?

What is the condition of man and animals during sleep?

Which of the functions go on less actively during sleep?

What is the amount of sleep necessary to a healthy person?

At what ages is the most sleep required? Why?

What effect has sleep on the pulse, on respiration, and digestion?

20

What is the state of the bloodvessels of the brain during sleep?

What is the act of dreaming?

What portions of the nervous system do animals possess in common with man?

Which portion does man possess in its highest form of development?

In the lowest forms of animal life, what evidences of a nervous system do we find?

How is the nervous system arranged in insects?

What are invertebrate animals?   Vertebrate?

In what part of the body do we find the nervous system in the invertebrate animals?   In the vertebrate?

How is the nervous system of the star-fish arranged? Of the mollusca?

What other peculiarities do insects present?

What peculiarity is there in birds?   In rabbits?

What portion of the brain becomes developed as we rise higher in the scale of animals?

What animals approach man in the general structure and shape of their brains?

What is the proportion which the brain of a monkey bears to that of man?

By way of summary of this chapter on the nervous system:—What are the two kinds of nerve-matter?   The functions or duties of each? What portions of the cerebrum are composed of white matter? Of gray matter? What portions of the spinal cord are made up of white matter? Of gray matter? Of which kind of matter are the nerves composed? What are the two great divisions of the nervous system called? What is included in the cerebro-spinal system? What is the arrangement of the sympathetic system? What parts are supplied by it? What general effect has it? What effect has it upon the bloodvessels? What are the functions of the cerebrum or brain proper? What are the functions of the cerebellum? What is the function of the medulla oblongata? What is the arrangement of the spinal nerves? What duties have the cranial nerves? What is the reflex action of the spinal cord, and where is it seated?   How are general impressions conveyed to the brain? What kind of movements are regulated by the reflex action of the spinal cord?   What are the functions of the cranial ganglia?

THE SENSES.

The Senses and their Objects.—Allusion was made several times, in describing the nervous system, to *general sensations*. It was there shown that they depended on the presence of a mind; that is, of a mind conscious of the impressions made on the nerves of sensation. We have now to consider *special sensations*, which are generally known as *the senses*, such as touch, taste, vision, hearing, and smelling. These all require special organs adapted to their perfection. Not only nerves are necessary, but also in some of them a physical arrangement, such as that of the eye with its delicate apparatus to receive the light, before the nerve is at all impressed by it. The object of the senses is to make us acquainted with the world around us. Without them we should be unable to appreciate the properties of matter, and would have no ideas of taste, odor, or sound, such as we acquire through the tongue, the nose, or the ear.

Cultivation of the Senses.—The five senses are not all as fully developed in man as in other animals. Man has his intellect more perfect, while in some of the other animals the senses, such as those of smell and hearing, seem superior to the powers of their intellects. Some of the senses, such as the taste, need cultivation in man, but they can be cultivated until they become sources of misery as well as of pleasure. The ear, for instance, can be trained to the utmost refinement of musical tones, but may be painfully annoyed by the slightest discord. So, too, the eye may be able to distinguish the faintest colors, and yet be constantly distressed by hues or tints that may not be blended in harmony. The most appetizing enjoyment of food and drinks may more than satisfy one's palate, and yet the taste may be cultivated to such an absurd extent that articles which please the fancies of almost any one else will be disgusting to his over-sensitive nature.

Distinction has been made between such senses as hearing and sight, which lead to actual exercise and gratification of the intellect, and are therefore called *intellectual senses*, and taste and smell, which do not subserve any such high purposes, but are merely *corpo'real senses;* that is, almost wholly for the enjoyment of the body, without exciting or leaving much intellectual pleasure in their train. The nerves connected with the senses are not capable of performing any other function except that for which they were created. The nerve of sight can do nothing more than fulfil the objects of vision; it does not, for example, give any evidence of pain; and it may be as truly said of the nerves of smell and of taste that they are incapable of any other functions than those of assisting in the appreciation of odors and flavors.

In addition to what are generally known as the five senses—touch, taste, smell, sight, and hearing—some physiologists describe hunger and thirst, the sense of temperature, and other sensations, as if they were special senses. From the earliest times, however, the five just mentioned have been the only ones universally recognized as "the senses," and as such it will be our province to describe them individually.

TASTE.

The Organ of Taste.—Taste is really a variety of the sense of touch, except that the surface is that of the tongue and the lining membrane of the mouth, which is a mucous membrane, and not, as in touch, the skin or outer covering of the body. Taste is not perfect unless this membrane is also perfect, for if the latter be injured in any way pain will take the place of taste. The actions of mastication and insalivation, already referred to in digestion of the food, which include the share performed by the teeth and salivary glands, enable the alimentary mass to be brought in contact with the nerves of taste in all parts of the cavity of the mouth. The sapid qualities of different articles are brought out by these very acts of division and moistening of the food. Taste is not properly accomplished, indeed, unless the action of the saliva and of the teeth is well executed. The impression is made on the nerves of taste, and thence conveyed to the brain, which appreciates it. While the chief organ of taste is the muscular organ known as the tongue and the mucous membrane covering it, other parts of the mouth contribute their share in perfecting the sense.

There are several important nerves which are sent to the tongue, but they do not all take part directly in the accomplishment of the sense of taste. The tongue is also supplied with muscles, which allow it to be moved in various directions, and also to take part in mastication or chewing of the food in swallowing. In the use of the voice certain letters of the alphabet are pronounced through the assistance of the tongue, as we shall see hereafter. Taste is under the control of the will, being exercised actively or passively, according to the degree to which the muscles of

20 *

the tongue are called into play. The organ of this sense is very properly placed at the entrance to the alimentary canal as an aid to the proper choice of food. Otherwise much of the food that is taken into the stomach might be rejected by that organ.

Papillæ of the Tongue.—When the surface of the tongue is examined it is found to be covered with an immense number of fine microscopic projections, called *papil'læ* or *villi*, some of which give it the smooth, velvety appearance it possesses, while others impart to it its roughness. It is in these papillæ that the delicate filaments of the nerve of taste are distributed. The papillæ are of different shapes, being simple like those on the surface of the skin, or thread-like (fil'iform), mushroom-like (fun'giform), and cup-shaped (cal'iciform). It is supposed that the two first named are concerned with the sense of touch of the tongue, the others with the sense of taste. These are distributed in all parts of the tongue. The tongue is also supplied with numerous little glands, which pour out a thin fluid to lubricate its surface and also the interior of the mouth. Indeed, the tongue must be moist and the article soluble before perfect taste can be accomplished. Experiments seem to indicate that the sense of taste exists over the whole surface of the back part of the tongue, the under surface of the tip, and in a narrow band along the edge of the tongue, and also in a portion of the palate or back and upper parts of the mouth. There is no sense of taste

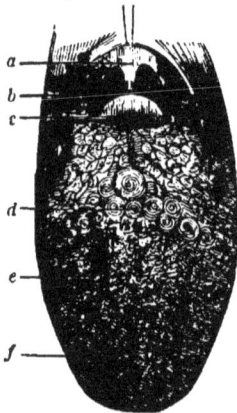

FIG. 91.—THE HUMAN TONGUE. (Showing also the back part of the mouth.) *a*, veil of the palate; *b*, tonsil; *c*, epiglottis; *d*, caliciform or cup-shaped papillæ; *e*, fungiform or mushroom-shaped papillæ; *f*, filiform or thread-like papillæ.

in the gums, in the inner surface of the lips, or in some other portions of the tongue. It seems that some parts of the tongue can appreciate salty articles while other parts appreciate sugars.

Taste.—We cannot define exactly the qualities which make some articles of diet agreeable to the taste, while others are disagreeable. In old times it was believed that the cause was due to one general principle in the various articles of food, which united with other elements to constitute the difference in taste between them. This is not the case, however, for they can be deprived of their taste or sapidity, as it is called, by cooking. Boiling water added to tea, for instance, separates the agreeable portions. What gives taste to an article, therefore, must be something peculiar to itself, which is different from that which gives flavor to another article. It was even thought at one time that sweet taste was produced by a little round granule, and sharpness or acidity by a pointed one; but this is not the case, for two articles of very similar shape have very opposite tastes, and they do not lose their individual tastes when in solution. Solution of a substance, as in water, separates the particles so that they may come directly in contact with the organ of taste; but a moderate degree of taste is not dependent on solution, for metals held in the mouth have a peculiar taste of their own.

Some physiologists have made the curious observation that sapid bodies have a different flavor according to the exact point on the tongue or the cavity of the mouth with which they are brought in contact. Alum, for instance, gives an acid and astringent taste when applied to the tip or extremity of the tongue, and a sweetish and not at all acid taste at the base of that organ; from which it has been inferred that the tip of the tongue has a special perception of sweet substances, and the base of the tongue of bitter ones.

Writers have divided savors into a variety of kinds, as sweet, sour, bitter, salty, etc., yet the simple division into agreeable and disagreeable is perhaps as convenient as any other, for as a rule—with important exceptions, however—substances of agreeable taste are generally useful, and those of disagreeable taste either injurious or without any advantage as articles of food.

While the sense of taste has its seat almost entirely in the mucous membrane covering the tongue, the other portions of the mouth—as already stated—are not wholly devoid of the feeling. Some cases in point have been mentioned by writers—one, for instance, in which a child was born without a tongue, and yet it knew that sugar was sweet and other substances bitter. Sometimes the papillæ of the tongue become completely saturated with the impression of an article that has been taken into the mouth, and a sort of after-taste lingers there, which prevents the taste of another article from being appreciated. This fact enables the medical man sometimes to give an agreeable substance to conceal the taste of one of a more disagreeable character with which he follows it. One who is blindfolded cannot, after a few moments, tell brandy or gin from wine or from one another if tasted in quick succession.

Quantity Necessary.—The quantity of matter necessary to impress the organ of taste varies according to the substance. An amount so small that it can scarcely be weighed on the smallest balance will sometimes produce a decided taste in the mouth. In the manufacture of beer the slightest amount of some very bitter material added during the process will penetrate a hogshead of the liquid and produce a very noticeable bitter taste. On the other hand, when

o water to impart sweetness to it, the solu-
or comparatively tasteless in any quantity
70 grains to a pint of water.

Tongue.—Three important nerves are dis-
tongue. The fifth pair of cranial nerves,
rve of general sensibility to the face and
, sends a *lingual* branch, which gives sensi-
ngue and greater perfection to the taste.
xperienced in that organ it is because deli-
f this lingual nerve are affected by some

The *glos'so-pharynge'al nerve* (literally,
the tongue and pharynx"), which is the
cranial nerves, also sends branches, and
iments have been made to prove that this
he special sense of taste. The two together
o perfect the sense. The *hypoglos'sal* nerve
ler the tongue"), distributed to the lower
gue, is a nerve of motion, which controls
ments of the tongue, swallowing, etc.

the Taste.—The sense of taste is not, like
hearing, one that can be educated or de-
y good or instructive effect it may have
It is more under the control of the will
sense, the muscles of the mouth being
nly at the wish of the individual himself.
acute in some persons than in others. It
l to a high state of refinement, as is shown
ne, who are able to distinguish with accu-
iar flavors and characteristics of vinous
taste, however, may be, so to speak, be-
rarily by indulgence or excess in drink-
not perfect unless the power of smell is
f the nostrils be tightly closed, the sense of

taste is blunted or destroyed. Many articles with powerful odors have a very slight amount of taste, but they are
rendered palatable by the very fact that they are odorous.
Those substances which are grateful both to taste and smell
may generally be considered as wholesome or likely to agree
with the individual, and those which are not pleasant both
to taste and smell as likely to disagree; but there are of
course exceptions to such a statement.

National Differences of Taste.—The impressions produced on the organ of taste are singularly variable. Much
depends on the healthy condition of the individual, and
sometimes on the emptiness or fulness of the stomach. In
Greenland, fish-oils and whale-oil, which would be repulsive to the tastes of Europeans or Americans, are largely
taken as food. Many articles of diet taken with *gusto* in
one part of the world cannot pass the sentinel of the organ
of taste in other parts. Some time in the last century,
when potatoes were first imported from America to France,
there was universal opposition to their introduction, although recommended by the royal favor of Louis XVI.;
their taste was considered detestable, and it was contended
that the most serious diseases would result from their
general use. Previous to the year 1860 horse-flesh was
looked on with almost universal disgust in France, but in
the siege of Paris in 1870–1 this repugnance was overcome
when the Parisians were compelled to resort to its use.
The taste then acquired has survived, and it is said that
there are now several hundred establishments devoted to
the slaughter of these animals and the preparation of this
kind of food.

Organ of Taste in other Animals.—In the higher classes
of animals the organ of taste is very much the same as in
man. The sense is not very strongly developed in birds,

in which the tongue is generally bony or cartilaginous in structure and the nervous papillæ are absent. The tongue of the woodpecker is sharp and divided like a fork, with which it pierces insects. The parrot uses his tongue to maintain a firm hold on the article he may be eating. In some reptiles the tongue is large and muscular. In serpents it is sharp and forked and very quick in its movements. In other animals, as the frog, it is projected with exceeding rapidity, to enable them to seize upon insects on which they feed. In the bee the tongue is like a little tube, which acts as a sucker through which it extracts juices from flowers. The tongue of fishes is often immovable in the throat, and perhaps covered with teeth.

QUESTIONS.

What are the five senses?

What is the general arrangement of the organs connected with them?

What is the object of the senses?

Are the senses more fully developed in man or animals?

What are the corporeal senses? The intellectual senses?

Do the nerves of the senses perform any other duties?

What other sensations are sometimes considered as special senses?

Of what sense is taste a variety?

What is the organ of the sense of taste?

How do mastication and insalivation assist in the taste?

What portion of the nervous system is the organ of appreciation of taste?

What effect have the muscles of the tongue?

How is the voice influenced by the tongue?

Is taste voluntary or involuntary?

What are the papillæ of the tongue?

What are the different shapes of the papillæ?

At what part of the tongue is taste appreciated?

What other parts of the mouth assist in the process?

What portions of the mouth are devoid of the power of taste?
What effect has solution on the taste or quality of food?
Into what classes may flavors be divided?
Is any fixed quantity necessary to produce a flavor?
What nerves are distributed to the tongue?
What is the function or duty of each?
Is the sense of taste capable of education?
What effect has the sense of smell on that of taste?
Are the same articles universally appreciated by the sense of taste?
What is the arrangement of the tongue or organ of taste in the higher classes of animals? In birds? In the woodpecker? In the parrot? In reptiles? In serpents? In the bee? In fishes?

SMELL.

The Organ of Smell.—The organ of smell is the lining or mucous membrane of the nose, particularly its back part. The object of the sense is to appreciate the odorous properties of bodies, which penetrate the interior of the nose in very fine particles. The delicate surface impressed contains a large number of nervous papillæ. The membrane is continued up into the interior of the bony cavities, called *na'sal fos'sæ,* which are covered in front by the nose proper, and open behind into the back part of the throat, so that the air can pass freely through. This mucous membrane is called the olfactory membrane, because it is the essential part of the organ of smell, or the Schneiderian membrane, after Schneider, a German anatomist, who first described it.

Each nostril is divided at its back part into three passages, which are so narrow that when they become swollen, as in a "cold in the head," they allow but little air to pass through, and therefore give rise to difficulty of breathing. Two of these cavities have communication with cavities in the bones, so that their extent is much increased. The

Schneiderian membrane, which is smooth and velvety, lines all the cavities, and pours out mucus to keep the parts moistened. This is largely increased in quantity, but altered in quality, in "cold in the head," forming a most annoying feature of that complaint. The tears also come down into the nose through a canal from the inner corner of the eye, and help to keep the membrane moist, as will be described under Vision.

The Nerve of Smell.—The nerve which is supplied to the nose as the nerve of smell is the first pair of cranial nerves, which comes off from the base of the brain. After it leaves that organ it divides into small branches, and passes through openings in one of the bones of the skull, and reaches the nasal fossæ, where they are scattered like a shower in fine branches over the upper and middle parts of the Schneiderian membrane (Fig. 92). To show what ignorance prevailed at one time in regard to the nerves, and this one particularly, it may be stated that anciently the olfactory nerve was supposed to be a canal through which the secretion of the nose was conveyed from the brain, in which organ the fluid was formed. The fifth pair of nerves—the nerve of general sensibility to the face —also sends its branches to the nose, and gives general sensibility to it, while it helps to perfect the sense of smell. The shape of the nose has probably but little to

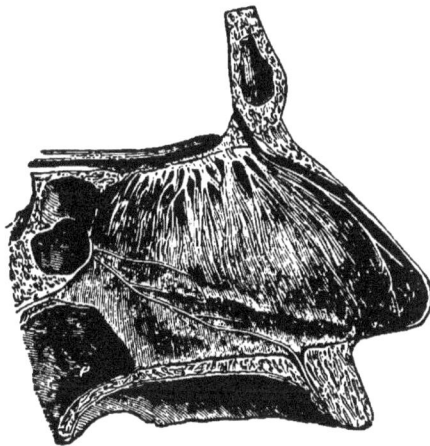

FIG. 92.—NERVES SUPPLIED TO THE NASAL FOSSÆ.

21   Q

do with the sense of smell, except that it supplies a sort of funnel in which to collect the odorous particles. The hairs at the extremity of that organ have been supposed to be put there to prevent unnecessary particles or insects from being carried into the nose with the air. As we shall see in the case of the eye and the ear, there is in the organ of the sense of smell an external arrangement to receive the impression and an internal one to appreciate it. The external or physical part in this case is the nose; the internal or nervous part is the olfactory nerve or nerve of smell.

Odors.—It is well to reflect for a moment on the nature of odors, which, like the particles that give taste to bodies, are emanations from various substances. Odors are part of the composition of these bodies, disengaged from them by heat, light, chemical action, moisture, etc. The effect of moisture in this way is shown after a summer's shower in a garden of flowering plants. Sometimes rubbing a substance may develop odor where it seemed to be absent previously. To enjoy a perfect sense of smell, the organ must be thoroughly open, the membrane natural, healthy, and moist, the nerve perfect, and the brain in a state of integrity. When the smell is temporarily destroyed, as in cold in the head, it is because the membrane is thickened and the nervous filaments covered, so that the odorous particles cannot reach a healthy surface.

The particles that convey odor to the organ of smell do not always require to be placed close to the nose. The smell of cinnamon has been detected at a distance of two hundred miles from land. Paper perfumed by a single grain of amber was once kept by a celebrated physiologist for forty years without any apparent loss of odor. The air is the usual medium for the diffusion of odors, but the art of the perfumer has enabled him to prepare and retain the most agreeable odors diffused in liquids.

It is just as difficult to classify odors as it was in the case of savors. So much depends on the nervous arrangement and development of the organs of smell and taste in each individual that no two persons would agree absolutely in a classification. They might, for convenience, be divided into animal odors (as musk), vegetable (as garlic), and mineral (as arsenic, when burnt). It is usual, however, to distinguish them, according to their resemblance to other substances, as garlicky, etc. If, however, we divide them, like the savors, into agreeable and disagreeable, we may run into the error of calling agreeable some that are exceedingly disagreeable to other persons.

How Smell is Effected.—The exact mode in which the sense of smell is exercised may be summed up in a few words. The air containing the odor is drawn into the nose, sometimes forcibly, so as to carry it as far as possible into the nasal cavities, in which the nerve of smell is distributed. The mucus poured out on the lining membrane helps to arrest the progress of the particles, so as to enable them to affect the olfactory nerve. The nose, from its funnel shape, collects the particles and directs them toward the inner parts.

Smell and Taste.—The senses of taste and smell aid each other, as already stated—the smell often warning us against putting an article into the mouth that may be repulsive to us; but the sense of smell is not so useful to man as to some other animals, which depend largely on its exercise in the selection of their food.

Uses of the Sense of Smell.—It has been stated by physiologists that we lose the delicacy of the sense of taste when the sense of smell is destroyed. Experiments have been made of tasting wine with the nostrils plugged and with the eyes blindfolded, and the difficulty of deciding by the taste alone has been most marked. Smell can hardly be called an intellectual sense, although sometimes educated for business purposes, as in the perfumer's art; it is rather a corporeal sense, for it gives but little information to improve the mind, no matter how well developed the sense may be. It must not be forgotten that another useful purpose served by the organ of the sense of smell —the nose and its cavities—is in the process of respiration. A channel is thus offered for the entrance of air to the lungs, and if there be anything offensive in the atmosphere the nose may sometimes act as a detective to warn the individual of impending danger to his health from such noxious influences. It is not by any means proven, however, that bad smells and the emanations that produce them are necessarily causes of disease, for many places with particularly offensive odors have never been known as centres of disease, and atmospheric air productive of the most dangerous and widespread maladies may be perfectly inodorous.

Chemical Action.—It is contended by some that, unless a stream of air containing oxygen passes into the cavities of the nostrils at the same time with the odorous particles, no smell will be produced; also, that if a current of carbonic acid gas accompanies the odor the effect is arrested. This would seem to show that there is a chemical action connected with the perfection of this sense, and that the presence of oxygen is absolutely necessary.

Sense of Smell in Animals.—Animals which have the greatest development of the olfactory nerves are those which possess the strongest instincts of scent. The size of the nostrils and interior bones of the nose must also be taken into account. Sharks will collect from a considerable distance around a dead body thrown into the ocean. Some of the higher classes of animals have a much greater development of the sense of smell than man. Many of them have a much larger relative amount of surface than man for the purpose of increasing the extent of the olfactory or Schneiderian membrane, such as additional cavities and bones. Some of the more timid animals have an acute sense of smell, as if to protect them against attack by their enemies. In birds the sense is not strongly developed, although sometimes the part of the brain from which the olfactory nerves proceed may be of considerable size. Birds generally use their eyes more than their organs of smell in the search for food. In reptiles the olfactory organs—the cavities and membrane—are not well developed. Fishes seem to have a tolerably complete arrangement for smell; they have nostrils opening outward, and sacs containing small branches from the olfactory lobe, but the water which they breathe cannot be renewed like the air of the atmosphere. Yet some of them have a muscular apparatus to keep the water in motion, and it is believed that a few of them, as the shark and the ray, have a power of scenting as well as smelling. In insects and in animals of the crab or lobster kind there is no perceptible apparatus for this sense, although it is believed that it exists.

The scent of animals, so noticeable in the pointer and bloodhound, depends on their power to detect by the organ of smell animal effluvia, as they are called, which are the emanations of odorous particles given off by animals. These are generally dense or heavy, and are therefore noticed near the ground, which will account for the position taken by scenting dogs, the nose being applied by them as low as possible. The same remark is true in re-

gard to decaying matter, the disagreeable odors of which
are scarcely perceptible in the upper stories of the house,
while on the ground floor they are at their strongest. The
emanations, whatever they may be, which produce disease
in those who lie on the earth in the swamps of tropical
regions are a familiar illustration in point, the people
guarding themselves from danger by sleeping in ham-
mocks suspended from trees. It is said that the Bay of
Naples is so foul that upper apartments in houses in that
city are more desirable residences than those lower down,
and rent at correspondingly higher rates.

QUESTIONS.

What is the organ of smell?
What is the object of the sense of smell?
What is the lining membrane of the nose called?
How is it arranged?
What is the arrangement of the cavities of the nose?
What membrane lines these cavities?
What communication is there between the eye and the nose?
What is the nerve of smell, and where does it come from?
What other nerve goes to the nose? What are its duties?
What effect has the shape of the nose?
What are the uses of the hairs inside the nose?
What is the external or physical arrangement of the organ of smell?
The internal arrangement?
What are odors?
In order that the sense of smell may be perfect, what organs must be
perfect?
Through what media are odors diffused?
Into what division may odors be classified?
What is the process by which odors are appreciated?
What is the use of the mucous secretion of the nose?
How do smell and taste aid one another?
Is smell an intellectual or a corporeal sense?

What useful purpose does smell serve in respiration?

What effect have oxygen and carbonic acid on the smell?

What parts of the organ of smell increase the sense of scent in animals?

What is the condition of this sense in other animals? In birds? In reptiles? In fishes? In insects?

What does the scent of animals depend upon?

Why does a scenting dog point low down to the ground?

What facts in regard to health or disease may we learn from these low emanations?

TOUCH.

All the Senses Modifications of Touch.—The five senses have all been considered by physiologists as modifications of the sense of touch. Smell and taste have been already shown, in considering the organs involved in their exercise, to be illustrations of this fact, for smell depends on the contact of odorous particles, and taste on the contact of sapid bodies. Vision and hearing will also be shown to be similarly caused by the contact of matter that cannot be weighed or handled any more than can the particles interested in the production of taste and smell.

The Organ of Touch.—The skin affords an organ in which are distributed myriads of little nervous points called *papil'læ*. These are more numerous and more developed wherever the sense of touch is most acute, as in the fingers. In the papillæ of that region are also small bodies of fibrous tissue, closely connected with the nerve, and called *touch bodies*, which are supposed to render the sense of touch more acute, just as a hard corn pressing on a nerve increases its sensibility. All portions of the skin are endowed with the sense of touch, but this sense is most perfect in the hands. Here the papillæ are said to number 20,000 to the square inch. To appreciate fully this organ of touch, it is necessary to understand clearly the arrangement of the skin itself.

The Skin.—The skin is intended as a protection to the delicate parts beneath it, and also as a means of getting rid of certain watery and fatty matters from the system through the agency of glands emptying upon its surface. The outer layer of the skin is called the *epider'mis* (from two Greek words meaning " upon the skin "). It is made up of cells differently shaped, and varying in thickness in different parts of the body. It is thick, for instance, where it is subject to pressure, as in the fingers or heels, and thin over the lips. The epidermis is not supplied with nerves or bloodvessels, and therefore whatever pain may be felt in the skin, or whatever bleeding may occur from it from injury, must arise from the next layer of the skin, which is largely supplied with both vessels and nerves. The epidermis is perforated by numerous openings or pores for the passage of hairs and the ducts or canals of the sweat-glands (Fig. 93).

FIG. 93.—SURFACE OF THE PALM OF THE HAND (slightly magnified).

1, 1, 1, openings of ducts of sweat-glands; 2, 2, 2, grooves between the papillæ of the skin.

The deeper layer of the skin is called variously the *cu'tis* (Latin for " skin "), *der'ma*, *co'rium* (" the hide ") or *true skin.* It is made up of areolar or cellular tissue and elastic fibrous tissue ; the elastic tissue being in excess in parts that require great elasticity, as in the armpit, and the areolar tissue predominating where resistance is required, as in the sole of the foot. It is thick, too, in the latter case, while in the eyelids it is very thin. Beneath the cutis is a layer of fat, which also varies greatly in amount in individuals, but which gives the figure its roundness and symmetrical appearance.

The Papillæ.—On the outer surface of the cutis are the *papil'læ,* to which reference has already been made. These are projections of varying size, the average length of which is about $\frac{1}{100}$th of an inch and breadth $\frac{1}{250}$th of an inch. They are well supplied with minute capillary vessels arranged in loops (Fig. 94). The bloodvessels, which are

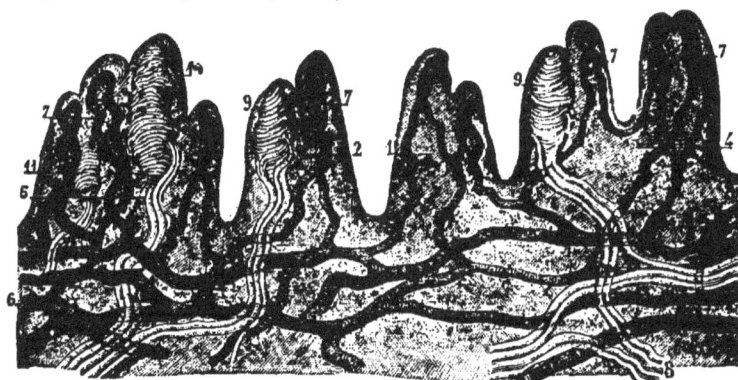

FIG. 94.—PAPILLÆ OF THE SKIN OF THE PALM OF THE HAND (greatly magnified).

1, papillæ with two loops of bloodvessels; 2, papillæ with a tactile corpuscle; 4, 5, large compound papillæ; 6, network of vessels beneath the papillæ; 7, 7, loops of vessels in the papillæ; 8, 8, nerves beneath the papillæ; 9, 9, 10, 11, tactile corpuscles.

derived from the arteries, are most numerous where the papillæ are most developed and the sense of touch most acute. If we take the hand as an illustration, we will find that these arterial and nervous loops are very much more numerous in the papillæ of the fingers or the back of the hand. The papillæ do not always contain nerves as well as bloodvessels; some of them have either one set or the other, but not both, but the most perfect of them possess both. Those papillæ which do not contain vessels are usually occupied by *tac'tile cor'puscles*, or *corpuscles of touch*. All these structures are represented in Figs. 94, 95, as seen under the microscope.

Sweat-Glands (Fig. 75).—The skin, as previously stated, offers a medium for getting rid of a large amount of fluid through the sweat or perspiration. This fluid is obtained from the blood through the agency of sweat-glands, or sudorif'erous glands, as they are technically called (from two Latin words meaning "sweat-producing"). Each gland is in the form of a tube convoluted, or turned upon itself, until it is in shape like a ball at its farthest extremity, and it has an outlet on the skin. The sweat-glands are very numerous, especially in the palm of the hand and the sole of the feet, where they are said to number nearly 3000 in each square inch of surface, or altogether in the whole body about 7,000,000. It is said that if all the sweat-glands in the body were placed end to end they would extend a distance of more than three miles. The sweat secreted by these glands is not generally visible; that is, it is carried off as a vapor, and is then called insensible perspiration. It is seen as a fluid when the amount is increased either by exercise or by the checking of the evaporation from the surface when the atmosphere is already loaded with moisture. It is then called sensible perspiration.

The amount of sweat secreted in twenty-four hours varies, but it has been estimated at two or three pounds in that period of time. Experiments made with animals have shown that if the perspiration was entirely checked by coating them with a varnish perfectly impervious to air, the temperature rapidly sank, and the animal died from interference with its respiration, the blood not being properly aërated in the lungs. The importance of this evaporation in the maintenance of animal heat has been already alluded to (p. 181).

Sebaceous Glands.—The skin is kept soft and pliable and is prevented from cracking or drying by the oily or fatty matter poured out upon its surface by numerous little glands, called *seba'ceous glands* or follicles (from the Latin word *sebum*, "suet"). They are abundant on the face and on the top of the head, in which last location they lubricate the hair; but they are not found on the palm of the hand or sole of the foot, both which surfaces are usually dry. They are more numerous in the inhabitants of warm climates, and in animals give off the odor which can be scented at a considerable distance by dogs.

Fig. 95.—Section of the Skin (as seen under the microscope).

1, epidermis; 2, dermis; 3, hair-follicles; 4, pigmentary layer; 5, corpuscles of touch; 6, sweat-glands; 7, fat-cells.

Some of the sebaceous secretion is also poured directly into the hair-follicles to anoint the hairs and keep them soft. The little dark points on the skin, of the face especially, from which "worms," as they are called, are pressed out, are only the ducts of the sebaceous glands obstructed so that they cannot empty themselves of their contents.

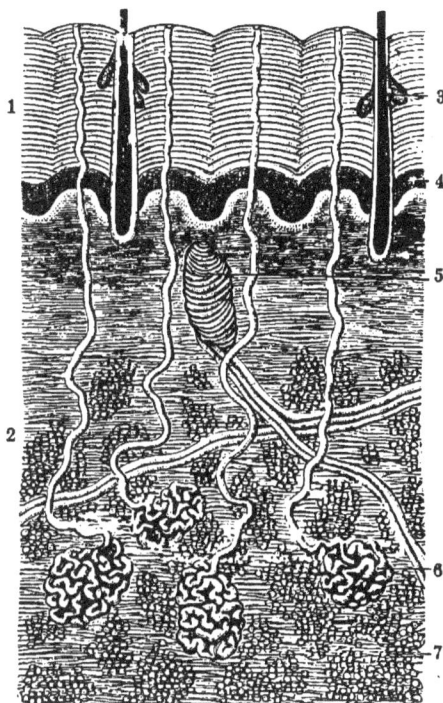

Hair (Fig. 95).—A hair is a much more complicated structure than would at first sight appear. It is divisible into a shaft and a root—one visible above the surface, the other having its origin in a follicle or sac within the skin. The shaft is made up of cells massed together, and differs in different animals. In man the shaft is solid and fibrous, with rough scales on its exterior. The shaft varies greatly in different animals, the scales being rougher in some than in others.

The outer extremity of the hair is naturally conical or pointed. When a hair is allowed to grow to any length it may split into fibres like a painter's brush, which will become brittle and break off. This is not so likely to happen if the sebaceous matter is sufficient in quantity to keep the hair from becoming too dry. The root of the thicker hairs passes deeply into the structure of the skin. It is rounder and larger than the shaft. It is shaped like a bulb at the bottom of the hair-follicle or sac.

The hair serves the useful purpose of protecting the surface of the body. It shields the head and face from danger of exposure to heat or cold. On the general surface of the body it doubtless helps to keep up the animal temperature. The eyebrows and the eyelashes certainly afford protection against dust or other injurious matters, and the hairs at the entrance of the nose and ears have a similar effect.

The *color of the hair* is dependent upon the presence of coloring-matter in the cells of the hair-follicles. It seems to be established that in very rare cases the color may be suddenly changed by fright or emotion, probably from the action of an acid secretion rapidly and suddenly formed, so that the statement of the poet, that " deadly fear can time outgo and blanch at once the hair," is not purely a fiction of the imagination. The *number of hairs*, which have been considered as practically beyond the power of man to count, has been estimated to be nearly 300 to a quarter of a square inch on the crown of the head of a man possessing an average amount of hair ; rather more than 200 over the forehead, and about 40 on the chin. A hair becomes gray or white from deficiency of secretion of the coloring-matter or pigment.

The *nails* are, like hairs, a turning in or inversion of the epidermis, forming a follicle or sac, to which blood-vessels are largely distributed. The fluid poured out becomes cellular, and afterward flat and condensed. Each nail is inserted in a fold of the skin, which is reflected backward to the root of the nail, and then passed forward beneath its under surface, to which it is adherent. They aid in the sense of touch, and also give a firm base of support for the extremities of the fingers. All the arrangements connected with the skin which animals possess for their comfort or protection, such as horns, feathers, hoofs, scales, bristles, etc., are varieties of the same general structure as the nails, only differing in the minor details of the mode in which they are constructed.

Color of the Skin.—The color depends on the deposit of a coloring-matter or pigment in the cells of the deeper layer of the epidermis. On its character and amount depends the difference of hue between the races. Black pigment is secreted in the negro, red in the Indian, etc. The differing fairness of skin in individuals of the same white race is connected with the amount of pigment so deposited. Freckles are nothing more than an increase of coloring-matter irregularly deposited. A person in whom the coloring-matter is entirely absent in the skin, hair, or the eye is called an albi′no (from *albus*, " white ").

22

The Hand.—This is in man the principal organ of touch. To assist in the exercise of this sense it is constructed in such a way as to be perfectly flexible, and by its variety of movements to adapt itself to the shape of the substance held within its grasp. In an instant it acquires appreciation of form, weight, consistence, etc. The joints between its numerous bones and the tendons which pass to every part of the hand (Fig. 96) admit of great variety of motion, and the thumb, which acts as a sort of opponent and at the same time an assistant of the other fingers, is a special and distinguishing characteristic

FIG. 96.—ANATOMY OF THE
HUMAN HAND (back part).

of man. The hand has always been regarded as one of the most valuable instruments which the Creator has given to the human race.

The Sensation of Touch.—The sensation differs with the character of the body touched, and with other agencies, such as the temperature. The sensation given by difference of temperature is not, however, to be considered purely a form of touch. Bodies of the same temperature, such as a stick or a stone, may give rise to greatly varying sensations of touch, or, as they are generally called, tactile impressions. Softness, hardness, smoothness, and roughness all enter into consideration as elements. The sensation endures for a little while after the removal of the body or substance that caused the sensation. If we touch two points on the skin at the same time, two different sensations

may result, but if the two parts touched are very close together there may be only one sensation. Some physiologists have, by means of the points of compasses applied
to various regions of the body, determined the existence of
the greatest diversity in their sensitiveness. At the tip
of the finger the double sensation was found to be experienced at the distance of only one-twentieth of an inch,
while in the middle of the back it was only perceptible
at a distance of rather more than two inches.

One would think that no deception could occur in regard
to our knowledge of the exact point that has been touched.
A familiar illusion, however, well known from times as far
back as the days of the old Greek philosophers, is the experiment of placing a marble between the fingers, as in
Fig. 97, and having the sensation of one marble imparted,
and afterward of crossing the index and middle fingers

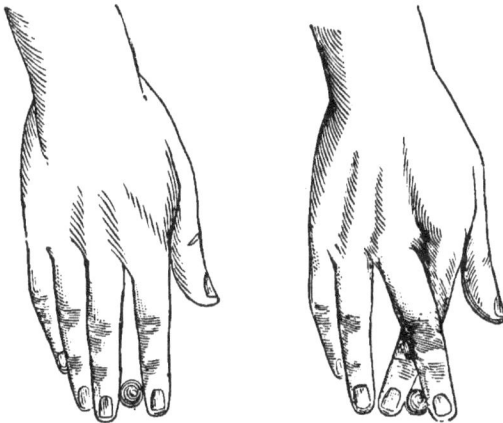

FIG. 97.—A TACTILE ILLUSION.

(Fig. 97), and having the sensation of two marbles. A
separate impression is here made on each finger in spite
of the knowledge that only one marble is placed there.

When the outer skin—the epidermis—is thick, the sense

of touch will be blunted, but quick and delicate when that coat is thin. If the epidermis did not exist touch would be replaced with the sensation of pain, and things that are now handled with impunity would be rapidly absorbed, and in many cases with poisonous effects. Like the other senses, touch may be educated or improved by proper exercise until it acquires the highest degree of refinement and delicacy, as is well seen in the execution of artistic workmanship. As has been well remarked, the educated touch of the surgeon is an instance of an acquired power of combining sensations of contact and of pressure, as indicated by the amount of resistance, and of forming a rapid and accurate judgment.

Sense of Touch in Other Animals.—This varies greatly and according to their needs. In most of the higher classes the tongue, lips, and snout are the chief organs of touch. In all of them it is placed near the surface as a guard and protection. In many of the insects the outer covering of the body is thick and shell-like, but they have near the mouth prolongations called feelers, which are very movable, and can be directed toward an object or toward the ground on which they walk. Bats possess these in a remarkable degree, and are aided by them in their rapid flights. The scales of fishes doubtless interfere greatly with their sense of touch. The whiskers of the cat and of other animals are generally in contact at their inner extremity with a branch of a nerve, and thus the impression derived from any body they may touch is communicated to it. In some experiments made with cats, blindfolded for the purpose, it was found that these animals could find their way readily out of labyrinths in which they had been placed, provided their whiskers were left intact, but that they were totally unable to do so when the whiskers were removed.

QUESTIONS.

Of what sense are the other senses a variety?
What is the organ of touch?
What are the papillæ?
What are the touch-bodies?
Is the sense of touch confined to any portion of the skin?
What is the use of the skin?
What fluids are separated through the skin?
What is the outer layer of the skin called? How is it arranged?
Is the epidermis supplied with vessels and nerves?
What are the openings or pores of the skin intended for?
What other names are given to the next layer of the skin?
Of what two kinds of tissue is it composed?
Which of these tissues is in excess in the sole of the foot?
What kind of tissue gives natural roundness to the figure?
Where do the papillæ arise? What is their size?
How are bloodvessels arranged in connection with them?
In what part of the hand are the papillæ most numerous?
If the papillæ do not contain bloodvessels, what other bodies take their place?
From what source is the sweat derived?
What is meant by the word "sudoriferous"?
What is the general arrangement of the sweat-glands?
In what parts of the body are they most numerous? How many are there in those regions?
How many such glands are there in the body?
What is the difference between sensible and insensible perspiration?
What is the amount of sweat secreted in twenty-four hours?
What is the effect of completely covering the surface of animals with varnish?
What are the sebaceous glands?
What kind of secretion do they pour out?
What effect have they upon the hair?
In what portions of the body are they absent?
Into what parts is a hair divisible?
What change takes place in a hair after it becomes long?
What useful purpose do the hairs serve?
What effect have the eyebrows and eyelashes?
What is the color of the hair dependent upon?
What does sudden blanching of the hair depend upon?

22 *   R

What is the number of hairs on different parts, as the crown of the head, the forehead, and chin?

Why does hair become white?

What are the nails? How are they formed?

What are their uses?

What other parts of animals are similarly formed?

What does the color of the skin depend upon?

What does the color of the races of mankind depend upon?

What are freckles? ·

What is an albino?

What portion of the body is the chief organ of touch?

What qualities has it for the purpose?

Which one of the five fingers is characteristic of man?

With what circumstances does the sensation of touch vary?

What is meant by tactile impressions?

By what experiments can we prove that some parts of the body are more impressible than others? ·

What deception in regard to sensation is there in the experiment with a marble?

What effect on touch has the thickness of the epidermis?

If the epidermis did not exist, what would be the result?

What are the chief organs of touch in the higher animals?

What are the feelers of insects?

What useful purpose do the whiskers of animals serve?

VISION.

Importance of the Sense.—There is perhaps no organ of the body which is the source of so much pleasure and instruction as the eye. To it we are indebted for a thousand enjoyments and advantages. By means of the vision we are able to appreciate light and color, and to recognize the form, size, movement, distance, and general properties of objects around us. The eye has always been a subject of curious study, for it is an organ founded on scientific principles, perfectly adapted for the duties it has to perform. A distinguished physicist has remarked that it is so perfect, so unspeakably perfect, that the searchers after tangible evidences of an all-wise and good Creator have declared their willingness to be limited to it alone, in the midst of millions, as their one triumphant proof.

The Organ of Vision.—The sense of vision is dependent on the optic nerve—which we have seen to be the second pair of cranial nerves coming from the base of the brain— and a complicated arrangement for the transmission of light to this nerve. In addition there are muscles which move the eye in various directions, and various parts which are intended chiefly for the protection of that organ. The parts essential to sight may be said to be contained in the *globe* or *ball* of the eye, besides the nerve of vision and the brain itself, both of which must also be perfect. It is not necessary for us to dwell at any length on the minute struc- ture of the eyeball, as it is a matter of careful anatomical study beyond the object of this work. It is composed of three coats or coverings—the first fibrous, the second vascular, and the third nervous—and of media that refract the light according to physical laws. The optic nerves as they pass from each side to the eyes cross one another like the lines in the letter X. This interchange of fibres of the nerves has something to do with their harmonious action, and probably with the sympathy existing between them both in health and disease.

Protection of the Eye.—The parts that act as guards and protectors of the eye are the eyelids, eyelashes, and eye- brows, with their thick skin and hairs, all of which shield it from excessive light and prevent dust from irritating it, while the tears and the sebaceous matter constantly poured over the front of the eye keep it moist and readily movable. These actions are examples of an involuntary kind of work, such as was described under the Nervous System as being controlled partly by the sympathetic system of nerves and partly by the reflex action of the cerebro-spinal system.

The Tears.—The tears are secreted by a gland called the *lach'rymal gland* (from *lacryma*, "a tear"), one on each side, seated above the outer part of each eyebrow (Figs. 98, 99). We are all familiar with the act of weeping, which is simply an excessive secretion of fluid from the lachrymal gland, usually under the influence of emotion, but we are not so familiar with the fact that the tears are also being secreted at every moment of our lives.

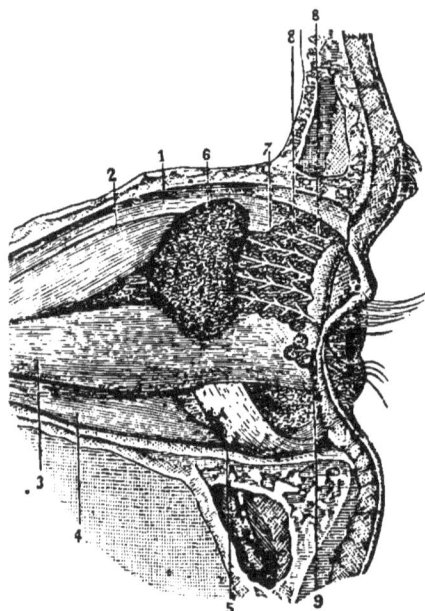

FIG. 98.—THE EYE. (The section shows the position of the lachrymal gland and of the muscles of the eyeball.)

1, 2, 3, 4, 5, muscles of the eyeball; 6, lachrymal gland; 7, 8, 9, ducts of the lachrymal gland.

The tears keep the surface clear and transparent. One can easily imagine how devoid of lustre the eyes would otherwise be. The fluid thus secreted or separated—the tears—is poured over the front of the eye by the movement of the eyelids, and afterward passes into the inner corner of the eye, into the opening of a little canal, which conveys it into the nose. The mouth of this canal can be seen as a little point in the corner of the eye next to the nose. The tears, being secreted at the outer part of the eye, and emptied at the inner corner into the nose, are

compelled to travel over the whole front of the eye. **The**
perspiration from the
forehead is prevented
from running into the
eye by the eyebrows
catching it and detain-
ing it.

The Orbits.—The eye-
ball is also protected
from injury by being
placed in a deep bony
cavity—the *socket*, as it is
generally called—which
is a hollow space pro-

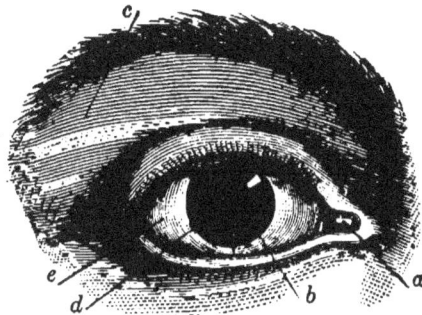

FIG. 99.—THE EYE.
a, opening of canal for passage of tears into the nose; b, iris; c, position of lachrymal gland, d, pupil; e, conjunctiva.

vided by nature for this purpose, partly in the frontal bone
and partly in the bones of the nose and cheek. The
frontal bone is that which forms the forehead in man
(Fig. 1). The term *orbit* is also applied to the socket.
The sides of the orbit project beyond the eyeball itself,
and in this way serve as a means of protection against
injury from force applied in any other direction than directly in front of the eye. There is a certain amount of
fat in the orbit which furnishes a cushion on which the
delicate eyeball rests. It is said that this bed of fatty
tissue is never entirely absent even in extreme emaciation.

The Eyelids.—The eyelids are like a curtain to the eye,
movable at will or involuntarily under the influence of appropriate muscles, the upper lid being larger and more free
in its motions. They are covered on the outside by skin,
but on their inside are lined by the same kind of membrane
which lines the cavities of the body—mucous membrane—
which is here, however, much more delicate, and is reflected
over the front of the eyeball. In this position it is called

the *conjuncti'va*, because it forms a kind of connection between the eyeball and the surrounding parts. It is a very sensitive membrane, and is frequently inflamed by the presence of dust or fine particles received into it from the air. It protects the eye, however, by its sensitiveness to such foreign bodies, for it makes a strong effort to get rid of them by bathing them with increased and sympathetic moisture of the tears and by the involuntary increase of the muscular action of the lids.

The Eyelashes.—The eyelashes collect the dust that might otherwise fall upon the delicate surface of the eye, and with

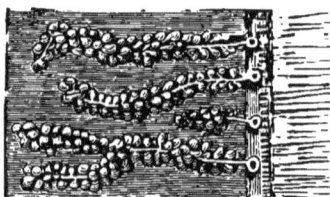

Fig. 100.—Glands of Meibomius (slightly magnified). The little openings back of the lashes are well shown.

the lids regulate the amount of light that would inevitably impinge with full force upon that organ. In close connection with the eyelashes are numerous little sebaceous glands (Figs, 99, 100), which pour out a thick oily fluid with which the edges of the lids are greased so that they do not adhere to one another. They also prevent the tears from overflowing the cheek instead of running into the eye to keep it moist. The arrangement of these little glands with their ducts is well shown in Fig. 100. They are called the *Meibo'mian glands*, after Meibomius, who first described them.

Transparent Parts of the Eye.—These may be clearly understood by reference to Fig. 101, which represents a section of the eyeball made in such a way as to exhibit clearly all the portions of which it is composed. The front and convex part of the eye, through which light must first pass, acts like a window to this organ, and is called the *cor'nea*, from its resemblance to horn (*cornu*, "horn"),

(Fig. 101, 1). Behind this is the *anterior chamber* (or front apartment) of the eye, as it is generally termed, containing a few drops of a watery fluid called the *aq'ueous hu'mor* (Fig. 101, 2). Behind this organ is a remarkable structure, the only solid part, called the *crys'talline lens* (Fig. 101, 5), which is convex on both its front and back surfaces and arranged in layers like an onion.

Farther back in the eye than the lens is another semi-

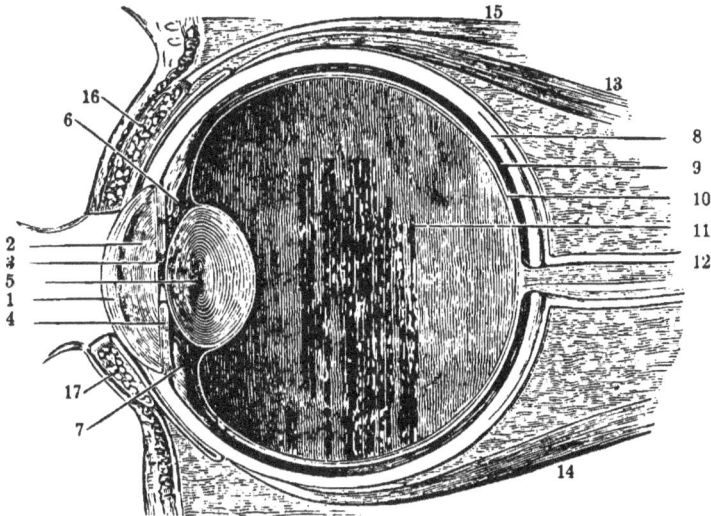

FIG. 101.—THE EYE. (A vertical section through the middle of the eyeball.)
1, cornea; 2, aqueous humor; 3, pupil; 4, iris; 5, lens; 6, ciliary processes; 7, canal around the lens; 8, sclerotic coat; 9, choroid; 10, retina; 11, vitreous humor; 12, optic nerve; 13, 14, straight muscles of the eyeball; 15, muscle that raises the upper eyelid; 16, upper eyelid; 17, lower eyelid.

fluid, transparent, gelatinous material, filling up the greater part of the eyeball, and called the *vit'reous humor*, because it resembles melted glass. This humor is placed in the *posterior chamber* (or back apartment) of the eye, and occupies about two-thirds of the eyeball. All these lenses or media for the transmission of light, whether they be horny, watery, gelatinous, or glassy, form when combined a beautiful optical apparatus without a single defect in its construction.

Coverings of the Eye.—The transparent portions of the eye, just described, are embraced in three distinct coverings or coats, placed one within the other. The outer coat, giving shape to the eye, is called the *sclerot'ic* (Fig. 101, 8), on account of its hardness (from a Greek word, meaning "hard"). This is a strong membrane, and supports the other delicate structures of the eye. The next is the *cho'roid* (Fig. 101, 9), a black, opaque membrane, with numerous vessels permeating it, and having pigment-cells connected with it to absorb unnecessary rays of light, and thus prevent confusion of vision. The third and last membrane, the *ret'ina* (Fig. 101, 10), lines the back part of the eyeball, and is placed between the choroid and· the vitreous humor. The retina is an expansion of the optic nerve, and is therefore the membrane which is sensitive to the action of light. In order to reach this membrane light must pass through the following transparent media: the cornea, aqueous humor, crystalline lens, and vitreous humor. In the front part of the choroid is a contractile structure, like a curtain, called the *i'ris* (Figs. 99, b, 101, 4), a movable diaphragm, with circular and radiating fibres, which dilate and contract the *pupil*, as the opening in the centre of the iris is called (Figs. 99, a, 103, 3). The pupil is the central opening, if we may so call it, of the eye, visible in looking into any one's eye while standing directly in front of him. The amount of light sent to the eye is regulated by this action of the pupil—or "apple of the eye," as it was formerly called—as all unnecessary light is shut out, and rays of light that strike on the eye outside the line of vision are arrested. This action is not under the control of the will, but is regulated by the reflex action of the nerves, according to the intensity and amount of light striking upon it. This is particularly noticeable in going from a dark to a light room, the pupil gradually contracting to accommodate itself to the light.

The *Cornea* (Fig. 101, 1) is quite a prominent feature of the front of the eye, and is in shape like the crystal or glass of a watch, and only about $\frac{1}{25}$th of an inch thick. It has no bloodvessels supplied to it. The *Sclerotic* is what is familiarly called the "white of the eye," an exceedingly strong membrane which gives attachment to the muscles outside the eyeball that control the movements of that organ. The optic nerve passes through an opening in its back part.

The *Choroid* is a very different membrane, being soft, delicate, and dark in color, and fully supplied with blood-vessels. Allusion has already been made to the dark layer of pigment-cells in the choroid coat, the uses of which are the absorption of rays of light after they have penetrated the transparent media of the eye. Telescopes of human construction are made on a similar plan, the interior of the tube being coated with some thick black material which absorbs superfluous rays of light, and thus prevents confusion. For a similar reason the eyesight of the albino, in whom this dark layer of the choroid is absent, is confused and imperfect, especially in the sunshine or in any other bright light. The iris (literally "a rainbow") is the part of the eye which gives to different persons the color—as blue, hazel, gray—that distinguishes them. The size of the pupil, or central opening of the iris, has much to do with the brilliancy of the eye, in young persons especially, for the light reflected from the front of the crystalline lens will then have more field for its display. The pupil is sometimes made to dilate artificially by the surgeon under the use of belladonna or some other drug, chiefly that he

23

may have a greater amount of space in which to apply his instruments.

The *Retina* is a thin, smooth, and delicate grayish or grayish-white membrane, made up of the filaments of the optic nerve. It is not sensitive to any other impressions than those of light. A blow on the head or over the eye, which would give rise to pain and suffering in the other coats of the eye, might only·produce flashes of light or confused images like sparks, as they are popularly called, from the retina. There is a point on the retina, just at the entrance of the optic nerve (Fig. 101) at the back of the eyeball, which is wholly insensible to the action of light. This is sometimes called the "blind spot" of the retina.

The *Crys'talline Lens* lies in the front part of the eye, behind the iris. It is enclosed in a membrane called a capsule, and is held in place partly by surrounding structures and partly by a delicate ligament. It is about a quarter of an inch thick, and is convex on both its front and back faces. Its position is well represented in Fig. 101. It is perfectly transparent, and when this transparency becomes lessened or lost, a condition called "cataract" results, especially in old persons, which often requires surgical relief.

The Motions of the Eyeball.—The muscles which move the eyeball are of course placed on the outside of that organ (Fig. 98). There are six of these muscles, four of which are straight and two oblique. Their action is confined to the rolling and turning of the ball; they do not ordinarily have any power to pull it inward or to project it outward. The line of vision is of course varied by such movements of the eyeball, and if necessary the head is also turned toward an object when the action of the muscles on the eyeball is not sufficient. Usually these movements of the eyeball take place without any consciousness of their action being aroused. Through the agency of these muscles the axis of vision can be turned in a horizontal plane of 60° toward the nose, and 90° outward, making a total of 150°. It can turn upward through an angle of 50°, and downward 70°, or 120° altogether.

Light.—The true cause of luminous sensations, as they are called, is the action of light on the retina, which is the expansion of the optic nerve. Light is known more from its effect than as being a tangible body which we can grasp and weigh, like bodies that are associated with the sense of touch. The theory which explains most fully the laws of light is that which supposes the existence of an elastic medium occupying all space, which is excited into rapid waves or undulations by the sun or any other luminous body. The theory is hence called the *un'dulatory theory* of light. The waves are propagated at an almost incomprehensible rate—nearly 190,000 miles a second being the estimated degree of rapidity. At this rate the light of the sun takes over eight minutes to reach us. The reception of these waves by the eye causes the sensation of light in that organ.

It will be found, in studying the principles of sound, so far as they are related to the organ of hearing, that a similar theory of sonorous vibration of the air is supposed to produce sound. This statement is sufficient for us, in the study of the physiology of the eye and ear, without entering more deeply into scientific explorations which do not at this time concern us.

Sight may therefore be said to be the effect of the movement of light on the retina, exciting the fibres of the optic nerve, the stimulus being conveyed to the brain for its appreciation. As a matter of curiosity, beyond our comprehension or our keenest conception, it may be briefly stated that at the lower end of the scale the sensation of red is produced by 435 billions of impulses or vibrations per second, and at the upper end of the scale the sensation of violet is produced by 764 billions of vibrations per second.

Reflection and Refraction of Light.—When light passes through a transparent medium of unvarying density, like the air, it proceeds rapidly in a straight line without being broken in its course. If, however, it comes in contact with any other transparent body of different density, it is either *reflected*—that is, turned back again in its course—or it passes through that medium in a different direction, being broken or *refracted*. The processes are called reflection of light and refraction of light.

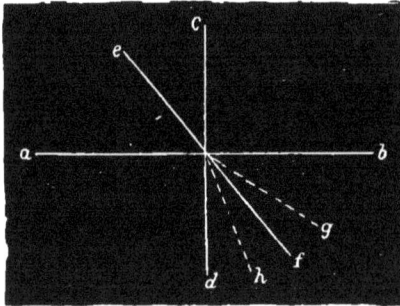

FIG. 102.—REFRACTION OF LIGHT.

If, for instance, *ab* be a transparent substance, as a pane of glass, a ray of light, *cd*, will pass directly through without undergoing any change. The ray *ef*, however, instead of passing directly through, will be bent in the direction *eh*. If the ray *ef* had passed from a dense medium into a rarer one it would have taken the direction *eg*. This represents the refraction of light, which becomes so important a consideration in the study of the physiology of the organ of vision.

When the rays of light strike upon the eye they pass in the first place through the cornea, which is convex, and therefore refracts them or turns them slightly from their course. They next pass through the aqueous humor, which is not sufficient to cause much change in the direction of the rays. They next strike upon the iris, which reflects or turns back those which impinge upon its surface, and only those rays which pass through the pupil, or central opening of the iris, reach the true organ of vision at the bottom of the eyeball.

It may be stated, in passing, that the *color of the eye* depends on the reflection of the light effected on the surface of the iris, or rather on the color of the choroid coat beyond the iris, seen through the latter. Dark eyes, for instance, are dependent on the deep color of the choroid as seen through the iris; light eyes, on a pale and colorless choroid similarly seen. When the light passes through the pupil the amount going to the retina may be diminished, as already stated, by the contraction of the iris diminishing the size of the pupil. This contraction in too intense a light, or the dilatation of the pupil when the light is not sufficiently bright, is dependent on a reflex sensation excited by the retina when impressed either with the excess or deficiency of light.

The rays of light, which we had traced as far as the pupil of the eye, next come in contact with the crystalline lens. A reference to Fig. 101 will show very clearly the regular order in which these organs are placed in the eye, and will enable the reader to follow closely the course of the rays of light in traversing that organ. The lens has decided refracting power, due to its double convex shape, resembling that of a small bean, and to the fact that the layers into which it may be divided increase in density toward the centre of the lens. The consequence is that the rays of light are still farther refracted from their course, and when they emerge from the lens and pass through the vitreous humor they converge on a small space on the

23 *

surface of the retina, producing an image upon it of the object.

The Image on the Retina.—By reference to Fig. 103 the convergence of the rays just alluded to will be understood, but it will be noticed also that the image impressed on the

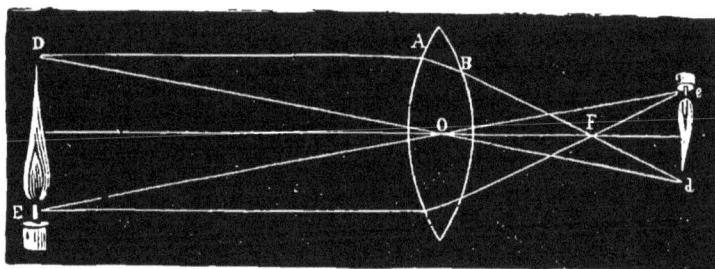

FIG. 103.—FORMATION OF AN IMAGE ON THE RETINA.

retina is reversed. The candle, instead of being in the natural upright position, as at DE, is seen at *ed* as it appears on the retina, with the flame downward. Notwithstanding

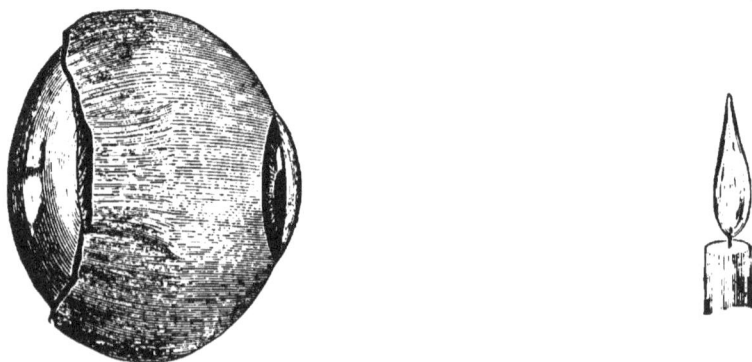

FIG. 104.—INVERTED IMAGE ON THE RETINA SHOWN IN THE BULLOCK'S EYE.

this inversion of the image, we see the object in its natural position. It has been supposed that the reason the image is not seen upside down is because during infancy the sense of sight is slowly and gradually educated, and that the child thus acquires a proper appreciation of the shape, size, and distance of objects. It is the brain, after all, and not the retina, which appreciates the existence of an image, although images must be constantly formed on the retina without the brain taking any note of them. Habit and comparison aid the brain in arriving at a proper conclusion as to the position of bodies at which the eye is looking.

Two Images Seen.—Each eye sees the same object at the same time; there must consequently be an image received by both eyes at once, and yet it is only appreciated as a single object. These two images are identical both in size and color and in every other particular; and when the retina transmits the two images of the same object to the optic nerve, and the optic nerve to the brain, the brain confounds the two in one and sees but one object. Perhaps the crossing of fibres of the two optic nerves, previously alluded to, may have something to do with the uniting of these two impressions.

Accommodation of the Eye to Distance.—As the eye is capable of seeing bodies at almost incalculable distances, some mechanism must exist for its adjustment for near and distant objects. In the natural condition of the healthy eye parallel rays (Fig. 105, A) coming from a distance are brought to a focus, or central point (x), on the retina. When the rays do not come accurately to a focus there, the image is quite indistinct, just as it is when an object is held so close to the eye that the refractive media of the eye cannot focus it on the retina. Other forms of eye exist in which no such accurate focusing takes place as in the healthy eye. In the *short-sighted* or *myop'ic* ("mouse-eyed") eye (Fig. 105, B) the parallel rays are brought to a focus (x) *in front* of the retina. To ensure distinct vision the object must be brought near the eye, so as to allow the divergent rays to be focused on the retina. In the *hypermetrop'ic* (from three Greek words meaning "beyond the measure of the eye"), or *far-sighted* eye (Fig. 105, c), the parallel rays of light are brought to a focus (*x*) *behind* the retina, so that it is necessary to hold the object far off in order to enable the less divergent rays to reach the eye. Different kinds of glasses must be used to correct the defects of the short-sighted and the far-sighted eye. The myopic or short-sighted eye may be corrected by using concave glasses, which will remove the focus farther back to the retina. The far-sighted eye (sometimes called the *presbyop'ic eye*, meaning "the eye of an old person") may be corrected by convex glasses, which will bring the focus farther forward.

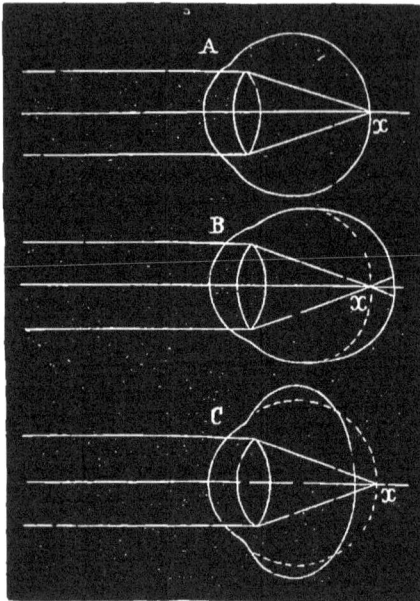

FIG. 105.

A, natural sight; B, short-sight; C, far-sight.

After a great deal of discussion on the subject, physiologists generally have come to the conclusion that the power of accommodation of the eye to distance is due to a change in the curvature of the front face of the crystalline lens. It is supposed that the lens in ordinary vision is flattened in front by the pressure of its capsule or covering, but that during the period of accommodation of the eye to distance the radiating fibres of a small muscle, called the ciliary muscle, pull on this capsule so as to relieve its tension, and the lens is then projected forward by its own elasticity.

Impression on the Retina.—An impression made on the retina is not always momentary, but may last for a while after the cause, as a light or bright object, has been removed. After the retina has been in the dark for a while it is more excitable, and of course after being exposed to a bright light for a while it is much less excitable. In other words, it is much more active after taking rest. An impression on the retina generally lasts from $\frac{1}{50}$th to $\frac{1}{30}$th of a second. An electric spark, which is itself of very short duration, may make so powerful an impression on the retina, and it will remain so long, that the spark will be visible. To show how persistent the impression is, it is only necessary to take a round piece of card having one side black and the other white, and rotate it, when only continuous dark bands will be seen. If a red spot be painted on the face of the disc it will appear, when the latter is rotated, like a red band, showing that the impression made by the spot is continuous. Each impression made upon the retina is here confounded with those which rapidly follow it. Another method of exhibiting the fact of this persistent sensation is by gazing steadily for a moment or two at a bright light, then looking at once into a dark room, when the image of the bright light will be distinctly seen. This shows that the original image has not yet faded from the retina. Several well-known toys are based upon this principle, one called the wheel of life, or the thaumatrope, being familiar to almost all. A ready mode of illustrating the same effect is in the rapid whirling of a stick lighted at one end, which appears like a ring of fire.

A peculiar effect of vision is what is known as *irra'diation*, which is shown in Fig. 106. The dark square in the white ground seems to be smaller than the white square in the dark ground, whereas they are of equal size. The only

S

explanation of this impression on the retina is that in vision the borders of the white surface advance and encroach on the dark ground. It is not easy to say why, but possibly

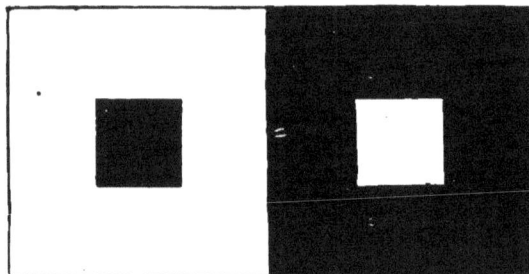

FIG. 106.—IRRADIATION.

a shadow which has the effect to diffuse the white surface forms around its edge, so as to make it appear larger.

There is one spot on the retina, called the *blind spot*, at which no sensibility to light occurs. This is at the entrance of the optic nerve, and there is also at the same point insensibility to color. This fact is illustrated by the following experiment (Fig. 107):

FIG. 107.—EXPERIMENT IN RE-GARD TO THE "BLIND SPOT."

Shut the left eye; with the right eye look steadily at the cross, and then move the book toward the eye and away from it. It will be found that the round spot will soon disappear, and it does so when the image from it falls directly on the entrance of the optic nerve. It is said that this blind spot, although small, is large enough to cause a human figure to disappear at a distance of about six and a half feet.

There is a *yellow spot* in the centre of the retina which is very sensitive to light. When we take up a book to read and run our eyes along the line, we do so with the view of bringing each word, as it comes in the text, opposite the yellow spot. This may be regarded as the centre of distinct vision, the field of which is not therefore very large. When we look at any objects whatever, their images are impressed on this limited field, and are rapidly mingled together, so that the brain appreciates them as a whole.

Perception of Colors.—In addition to the appreciation of light, the eye has also another duty to perform in the perception of color. There are seven elementary colors— violet, indigo, blue, green, yellow, orange, and red (remembered by the word "*vibgyor*," in which the first letter of each color is found). How early in life this sensation of color occurs is not definitely known, as we can learn nothing from the lips of infancy as to its personal experience. The blind from birth, when restored to light at an adult period of existence, require time before they can appreciate distinction of color. Some persons cannot distinguish colors at all, but only light and shadow. This rarely happens, however, the principal form of *color-blindness* being an inability to tell one color from another. Red and green seem to be the colors most frequently confounded. Red, blue, and yellow are mistaken for green, purple, orange, etc. The risk to railway travellers of employés overlooking the red signal of danger, owing to some visual defect in the perception of colors, is too serious a matter to be lightly passed over. In all parts of the world special attention is now being paid to the subject, and the eyes of the railway officials are being systematically tested. It is difficult to say wherein the defect lies, whether in the optic nerve or the retina or the different media through which the light travels. If, for instance, the humors of the eye themselves have even a slight color, some parts of the light will be intercepted before they reach the retina, and the image on the retina will therefore be deprived of the colors which are intercepted. Suppose, for instance, from any cause the humors of the eye should have the power of intercepting all the blue and violet rays of white light; all white objects would appear on the retina as if they had a reddish color.

Stereoscopic Vision.—When we look at an object with both eyes the image falls at the same time on the yellow spot of each eye, and we see but one object. If we press on one eye with the finger we see two images, not equally distinct, because the image will not fall on the yellow spot of that eye. We see only one image, indeed, if the light falls on corresponding points in each eye, whether it be the yellow spot or not. If, however, we look at an object, first with one eye and then with the other, it will be found that the images produced are different. By using a stereoscope, an instrument devised for the purpose in which the glasses correspond with the angle of the two eyes, the two images are combined and apparent solidity is imparted to them. The term *stereoscop'ic* has been applied to this kind of vision, from two Greek words meaning " to see a solid."

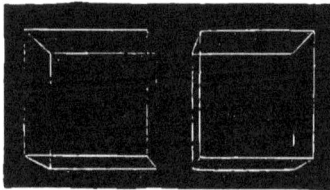

FIG. 108.—STEREOSCOPIC VISION.

Vision in the Inferior Animals.—A great variety exists in the different classes of animals in the construction of the organs of vision. In some of the lower forms of animal life the eyes are scarcely anything more than dots capable only of distinguishing light from darkness.

In vertebrate animals—those which have a spinal col- umn—there are two eyes, spherical shaped, resting in a socket lined with fat as a cushion, just as is the case in man. In some fishes, and in some of the lower forms of animal life that live both in land and water, the eyelids are absent and the eyes are covered with a skin. In the chameleon there is a disc or circle of skin with a central opening in it, and in some other animals there are not only eyelids, but also a very movable third membrane called the nic'titating membrane. A familiar illustration of this membrane is seen in the tortoise. In many birds it can be drawn over the whole surface of the eye to protect the organ from strong light or from foreign bodies. In the horse there is a triangular-shaped cartilage, called the haw, which can be drawn closely over the convex surface of the eye for the removal of dust or insects. Some animals, as fishes, have no tear-apparatus.

The sclerotic coat is hard in man, but in some animals, as birds and lizards, it is composed of cartilage, if not of bone. In fishes the eye might suffer by concussion of the water against it but for some such arrangement. In some animals the choroid or its layer of pigment—which we have seen, if present, absorbs the rays of light—is absent, so that the eye looks white or metallic. In the cat and the lion a portion of the choroid coat is covered with a bluish layer with a metallic lustre; the light is therefore strongly reflected, so that the eyes seem like balls of fire. The pupil is not always round; it is sometimes elongated vertically, as in cats and crocodiles; sometimes angular, as in some toads. In the owl the pupil is much more movable than in man. In nocturnal animals—that is, those who seek their prey by night—the eyes are usually larger than in those which feed during the day; in the daytime, too, the pupil becomes elliptical, appearing almost like a buttonhole, while at night it is circular, as in man.

Insects and the crustacea, which last class includes crabs and lobsters, have *compound eyes,* as they are called, or eyes made up of an immense number of conical tubes, which go

24

off like rays from a common centre, like spokes of a
wheel placed very closely together, and form a spherical-
shaped surface on the exterior of the eye. Each ray
has a sort of cornea or window at its extremity; in its
interior a humor like the vitreous humor of the human
eye, and at its inner end a filament of a nerve, the inside
of the ray being coated with pigment. All these filaments
or threads of nerves unite in a single optic nerve. There
are sometimes from ten to twenty thousand of these tubes
in a single eye, each one of which is really an eye itself, and
all of them packed away in a space of a minute fraction
of an inch. As a matter of interest to the curious, it may
be stated that the ant has 50 of these eyes, the common
fly 4000, the silkworm more than 6000, the goat-moth
more than 11,000, the butterfly more than 17,000, some
forms of beetle about 25,000. Insects so furnished must
be provided with eyes for every direction.

A few other peculiarities of vision in animals generally
may be briefly stated. Eyes are not always placed, like the
human eye, symmetrically; that is, similarly on opposite
sides of the body. Spiders, for instance, have from half a
dozen to a dozen eyes placed on a prominent part of the
back, instead of on the sides of the head. Some animals,
such as those insects which exist only in dark caverns or
in deep wells, to which the light never penetrates, are entirely devoid of organs of sight. Birds are far-sighted, or
presbyopic (see p. 271), the lens being more flattened than
in man. Fishes, on the other hand, are near-sighted, or
myopic. As the water in which they live is a denser medium than the air, a greater amount of refraction takes
place; for this purpose the crystalline lens is more convex
than in man.

QUESTIONS.

What useful purpose does vision serve?

What is the nerve of vision?

Which pair of cranial nerves is it?

In what are the parts necessary to sight contained?

Of how many coats or coverings is the eye composed?

What is the course of the fibres of the two optic nerves?

What parts protect the eye?

What effect have the tears and sebaceous matter?

What systems of nerves control these secretions and the movements of the eyelids and eyelashes?

What gland secretes the tears?

When does the secretion take place?

What other uses have the tears?

What finally becomes of the tears?

What prevents the perspiration from running into the eye?

What is the socket of the eye? The orbit?

What is the nature of the protection given the eye by the socket?

What is the cornea?

Of what use is the fatty matter in the orbit?

What are the eyelids, and how are they arranged?

How are they covered and lined?

What is the conjunctiva?

How does it protect the eye?

Of what use are the eyelashes?

How are the edges of the eyelids greased?

What other useful service does this secretion serve?

What is the name of the glands which secrete them?

What is the anterior chamber of the eye, and what does it contain?

What is the quantity of the aqueous humor?

What is the crystalline lens?

What is the fluid in the posterior chamber of the eye called?

How much of the eyeball does it occupy?

What is the object of these transparent media?

What are the three coats of the eye called?

How is the choroid coat arranged to absorb the rays of light?

What is the nervous coat of the eye called?

Of what nerve is it an expansion?

Through what transparent media must the light pass to reach the

What is the iris?

What is the opening in its centre called?

What is the apple of the eye?

What useful purpose does the pupil serve?

What system of nerves controls this action?

What is the shape of the cornea?   Its thickness?

What is the white of the eye?

What muscles are attached to the sclerotic coat?

What nerve passes through the back part of this coat?

What kind of a membrane is the choroid?

Which one of the coats of the eye is most largely supplied with blood-vessels?

What is the purpose of the pigmentary layer of the choroid?

How are telescopes arranged in a similar way?

What is an albino?

What part of the eye gives the color to the eyes of different indi-viduals?

Does the size of the pupil give brilliancy or the reverse to the eye?

How may the pupil be dilated artificially?

What kind of a membrane is the retina?

To what kind of impressions is it sensitive?

Is the whole of the retina sensitive to light?

What is the blind spot of the retina?

In what is the crystalline lens enclosed?

How thick is the lens?   What is its shape?

What is cataract?

How is the eyeball moved?

How many muscles are connected with the ball?

What kinds are they?   What action have they?

Through what number of degrees can the axis of vision be turned inward?   Outward?   Upward and downward?

What is the cause of luminous sensations?

What is the correct theory of light?

Why is it called the undulatory theory?

What is the rate at which light travels?

Can sound be explained on similar principles?

What is the definition of sight on such an explanation?

What is the estimate of the number of vibrations necessary to pro-duce red and violet?

When does light undergo change in passing through media?

What is reflection of light?

What is refraction of light?

When light strikes upon the eye, what does it first pass through?

What effect has the cornea upon it?

What effect has the iris upon it?

What does the color of the eye depend upon?

What are dark eyes dependent upon? Light eyes?

What influence has the pupil in controlling the amount of light?

What nervous action is this dependent upon?

What is the cause of the refractive power of the lens?

What is the position of the image on the retina?

Why do we not see it upside down?

What becomes of the two impressions made, one on each eye?

What becomes of the parallel rays of light in the healthy eye?

What becomes of these rays in the near-sighted eye? In the far-sighted eye?

What is a myopic eye?

What is a hypermetropic eye? A presbyopic eye?

How can the short-sighted eye be corrected?

How can the far-sighted eye be corrected?

To what is the power of accommodation of the eye due?

How does this curvature of the lens take place?

How long does an impression on the retina last?

What effect has rest on the excitability of the retina?

By what experiments can we prove that persistent impressions can be made on the retina?

What is irradiation? How is it explained?

What is the blind spot of the retina?

What effect has it on vision?

What is the yellow spot?

What effect has it on vision?

What are the seven elementary colors?

How early in life does this appreciation of color occur?

What is color-blindness?

What colors are generally confounded?

What is the cause of the defect?

What effect might the humors of the eye have in causing it?

On what spot of the retina must images fall to ensure perfect vision?

What is stereoscopic vision?

What peculiarities of the eyelids are noticed in fishes?

What third membrane exists in the tortoise? What is its object?

What cartilaginous arrangement is there in the horse's eye?

24 *

What peculiarities of the sclerotic coat are found in birds and lizards?
What changes occur in some animals in the layer of pigment of the choroid coat?
Is the pupil always round, as in man?
What peculiarities are noticed in the eyes of nocturnal animals?
What are compound eyes?
Describe the arrangement for vision in such eyes.
How many of these tubules are usually met with?
How many has the common fly? The butterfly?
Are eyes always placed opposite, as in man?
Are there any animals wholly devoid of eyes?
How does the crystalline lens in birds compare in shape with that of man?
How does the shape in fishes compare with that of man?
Why are birds presbyopic and fishes myopic?

HEARING.

The Apparatus of Hearing.—The organ of hearing is the ear. It consists of three distinct portions—the outer or external ear, the middle ear, and the internal ear, in which the nerve of hearing is distributed.

The *External Ear* (Fig. 109) includes the *concha* ("shell")—which is the technical name for the projecting organ on the side of the head familiarly known as the ear—and a canal (Fig. 109, *a*), leading inward, which is called the *external aud'itory canal* or *tube*. This tube is closed at its inner extremity by a membrane called the *membrane of the tym'panum* or drum, having the appearance in Fig. 109 of a radiated disc.

The *Middle Ear* is a cavity filled with air, separated from the external ear by the membrane of the tympanum, and communicating with the back of the throat by a tube (Fig. 109, *b*), the *Eustach'ian tube* (pron. Eus*take*ian) or canal, named after the anatomist Eustachius, who first described it. The membrane of the tympanum is connected with an opening called the *oval window* by a chain of little bones (Fig. 109, *d, e, f*).

The *Internal Ear* is called the *lab'yrinth*, because it is the most intricate part of the organ of hearing. It is

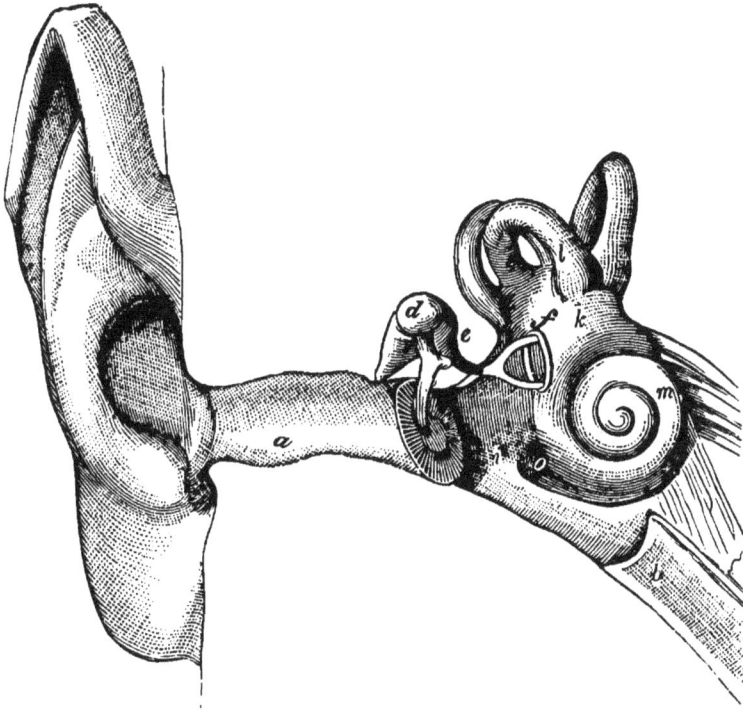

FIG. 109.—THE EAR.

(The different parts of the ear are here shown, divested of surrounding bony matter.) *a*, external auditory canal or tube; *b*, Eustachian tube; *f*, oval window; *o*, round window; *d, e, f*, little bones of the ear; *i*, middle ear; *k*, vestibule; *l*, semicircular canals; *m*, cochlea. (All the inner parts are here larger than natural.)

filled with fluid, and consists of three important parts— the vestibule, the semicircular canals, and the cochlea (Fig. 109, *k, l, m*). Each consists of a bony and a membranous portion. The internal ear is the absolutely essential portion of the organ of hearing, in which the auditory nerve—the nerve of hearing—is distributed. This nerve is one of the seventh pair of cranial nerves. The vestibule exists in every class of animals in which an apparatus for hearing has been detected. The semicircular canals are on one side, the cochlea on the other, and the vestibule lies between the two. In the wall of the vestibule is an opening called the oval window (Fig. 109, *f*), closed by a

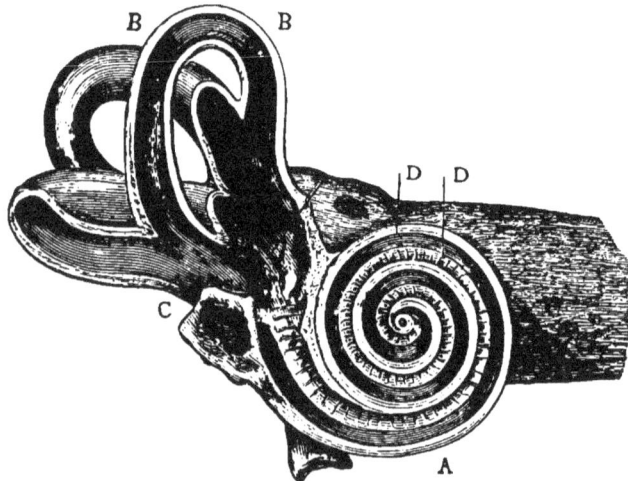

FIG. 110.—INTERIOR OF THE INTERNAL EAR.

A, D, cochlea; B, B, semicircular canals; C, vestibule.

membrane, and one of the little bones of the ear is inserted into it, as shown at *f*, Fig. 109. The *little bones of the ear* are called, from their fancied shape, the hammer, the anvil, and the stirrup. Their likeness to these articles may be seen in Fig 109, *d, e, f.* The vestibule has seven openings communicating with it, and it contains a large amount of earthy matter, either hard, like little stones—to which the name *o'toliths* ("ear-stones") has been given—or fine, like dust or powder ("ear-sand"). These doubtless have the effect of adding to the force and intensity of the sound. Hair-like bodies are noticed on the walls of the vestibule, in contact with cells connected with the auditory nerve.

When we examine the labyrinth as a whole we find it made up of apartments or chambers and canals which have been naturally hollowed out in the temporal bone (Fig. 111). The osseous or *bony labyrinth* is nothing more than a mould in the hardest part of that bone, containing a small quantity of fluid, the *per'ilymph* (literally, "the fluid around it"). In this mould, and surrounded by this fluid, is what is called the *mem'-branous labyrinth*, in which organ is the nervous part of the organ of hearing, bathed

FIG. 111.—THE COCHLEA (representing its spiral structure).

in another fluid, the *en'dolymph* (literally, "the fluid within it"). The membranous labyrinth consists partly of the vestibule communicating with the semicircular canals by five distinct openings. Each canal consists of a tube, bulging at each end, in which there are cells with long hairs. These hairs are the ends of filaments of the auditory nerve, on which impression must be made in order to produce hearing. The other division of the internal ear or labyrinth is called the *coch'lea* ("a snail-shell"), (Figs. 109, *m*; 110, A, D). This is a continuous triangular tube, turned in a spiral manner, and divided internally in its length, by an incomplete partition, into two parts, the lower of which extends to the round window, and the other goes directly to the vestibule. This spiral arrangement—called the *scal'æ*, or staircases—is well shown in Fig. 111.

Functions of the External Ear.—The external ear, sometimes called the *auricle*, has no other duty connected with hearing than to collect the sonorous waves and transmit them through the canal to the membrane of the drum of the ear. The familiar habit of applying the hand behind the ear when we desire to increase our capacity for hearing is practised with the view of collecting the vibrations or waves of sound, so that they will pass in greater quantity and force into the external ear. Some animals lower in the scale have very long ears or appendages externally, and it has been observed that the loss of such parts is frequently in them a cause of deafness. In some the auricle is trumpet-shaped, and is freely movable, so that sound can be collected from all directions. The external auditory canal (Fig. 109) is protected from injury from insects and foreign bodies by a form of secretion called ear-wax, poured out by certain glands called *ceru'minous glands* (from *cerumen,* "wax") on its surface, and by the hairs that grow along the canal. Even the elevations and depressions on the outer ear assist in the propagation of sound, and if these be filled up with some soft material, so as to make the surface even, the individual cannot hear quite as distinctly, or tell in precisely what direction the sound is coming.

Functions of the Middle Ear.—The middle ear—the drum or tympanum (Latin for "drum"), as it is generally called—is chiefly remarkable for the presence of the small bones previously alluded to. These bones are moved by very small muscles, and they have the effect, by their motions, of rendering the membranes of the tympanum with which they are in contact either tight or loose, according to the intensity of the vibrations of sound that fall upon them. The cavity of the drum of the ear being filled with air passing from it to the back of the throat, the little bones are allowed free vibration, and the air in the middle ear is thus maintained at a uniform temperature.

Another good effect upon the membrane from the presence of air here is from the atmospheric pressure being made equal on both sides of the membrane of the tympanum or drum. It is generally supposed that the Eustachian tube is closed during the act of swallowing food or drink, and that it is open during repose.

The chief muscle connected with the membrane of the tympanum is called the *ten'sor tym'pani* (literally, " tightener of the drum "), which pulls the membrane inward toward the cavity. After it has acted, the membrane returns to its natural relaxed condition, owing to its own elasticity and that of the chain of bones connected with it. By means of such movements the membrane is able to accommodate itself to the receiving and transmitting of sounds of varying pitch—to high sounds when it is tense, to low sounds when it is relaxed. When the membrane of the tympanum vibrates, the vibrations are transmitted to the internal ear, partly through the air in the middle ear and partly along the solid bones—ossicles (" little bones "), as they are called—the anvil, hammer, and stirrup. The power of solid bodies, as of liquids, to convey sound is well understood and familiar to everybody. These bones weigh only a few grains.

Functions of the Internal Ear.—In the internal ear sound may reach the labyrinth through its bony walls by means of the air in the drum of the ear striking upon the round window, or through the stirrup-bone which is attached to the oval window. Sounds may be distinctly heard that have no origin or cause outside the head itself, but which may be due to the action of muscles or the passage of blood along the bloodvessels in close relation to the nervous portion of the ear. Hearing is probably effected usually through the pulsation or vibration of the little bones, communicated

to the fluid in the labyrinth, and thence to the minute terminations of the auditory nerve.

That hearing may be accomplished it is necessary that
oscillations in the air, in water, and in solid bodies, which
give rise to the sensation of sound, should reach the filaments of the auditory nerve, and that these should be con-'
veyed to the brain for its appreciation. It will be seen,
therefore, how necessary it is that both the brain and the
auditory nerve are in a healthy condition. The vibrations of sound when they fall upon the ear come in contact
with the vestibule in which is the auditory nerve. The
whole organ of hearing in the internal ear is so delicate
and sensitive that the slightest vibration communicated to
it, and especially to the fluid contained in it, must act upon
the special nerve of hearing. The vibrations of sound may
reach the labyrinth through the external ear or through the
bones of the head. The latter channel may be proved to
exist by placing a watch between the teeth, when its tick-
ing will be distinctly heard. The sound of one's own voice
is heard in the same way.

Sound.—In order that sound may be fully appreciated a
certain amount of fulness is necessary to it. If too feeble,
no such result will occur; in other words, hearing will not
take place. It has been stated by some physiologists, with
a view of expressing the lowest limit capable of effecting
sound that will be heard, that it is represented by a pith-
ball, about $\frac{1}{65}$ of a grain in weight, falling upon a smooth
piece of glass from a height of $\frac{1}{20}$ of an inch at a distance
of $3\frac{1}{2}$ inches from the ear. A certain number of vibrations
is necessary to produce a tone, otherwise sound will not be
heard. The lowest limit is said to be about 30 vibrations
a second, the highest between 30,000 and 35,000. *Intensity*
of sound must be distinguished from pitch. Intensity de-

pends on the greater or less degree of fulness of the vibrations. *Pitch* depends on the number of vibrations in a
particular period of time, or, to phrase it differently, on
the length of time occupied by a single vibration. *Quality*
of sound is another element of consideration. It depends
on the form of the wave of sound that reaches the ear.
By this appreciation of the quality of sound we are able
to distinguish the voices of individuals, the sound of
special musical instruments, etc.

A sensation of sound lasts, as we have already seen to be
the case with a sensation of light, after the cause has been
removed. We are all familiar with such experience. The
two ears act in harmony, and hear but a single sound, just
as the two eyes see but a single image, and the brain in the
two instances appreciates but one object and one sound. The
eye and the ear have many points in common. The theory
of sound, as stated under Vision, is based on a similar idea
to that of the transmission of light; that is, by waves or
undulations. There is a physical apparatus—the outer
portion of the ear—just as there was a physical arrangement in the eye, before the nerve was acted upon by waves
of light or of sound. The intensity of sound, as of light,
is regulated in its action on the ear by a muscular apparatus
not under the control of the will.

The Sense of Hearing in Other Animals.—All the higher
classes of animals possess external ears, which in some of
them are capable of being turned in various directions to
catch the sound. In some animals, as the bat, they are out
of all proportion to their whole size. In some animals the
outer ear consists of several pieces, instead of one piece, as
in man. The interior ear does not differ much in different
animals. Curves of all kinds characterize the semicircular
canals, some being elliptical, others a portion of a circle,

25   T

etc. While in man the cochlea takes two and a half turns, it makes nearly four turns in the squirrel and only one and a half in the whale. Many of the higher classes of animals have the ear lodged in a separate bone—the tympan'ic bone—and not in the temporal bone as in man. Birds generally are devoid of an exterior ear, but in some of them the drum of the ear communicates with cavities in the bones of the skull, which act like a sounding-board of a piano to increase the resonance, as it is called. In reptiles the organ of hearing is imperfect, the external ear and auditory canal being sometimes absent, but they possess a drum and its membrane, the little bones, and a cochlea. The little ear-stones in the internal ear are especially noticeable in fishes, some of which possess, as their whole organ of hearing, a sacful of such bodies, taking the place of the cochlea, and having the nerve of hearing distributed on its walls. There is often no other part of the outer ear present except the external auditory canal; the cochlea, too, is straight and not coiled as in man. Sometimes there is only one semicircular canal. In the soft forms of animal vegetation called the mollusca the vestibule is the only part of the internal ear that is present; in fact, as previously stated, this part of the labyrinth is never absent in any animal with a defined apparatus for hearing. It contains oscillating bodies which resemble the otoliths or ear-stones previously described. In insects the organ of hearing is said to be wholly deficient, but it is possible that it may sometimes exist where not suspected, as they produce various forms of noise by which they call and answer one another.

QUESTIONS.

Of what three parts does the organ of hearing consist?
To which part is the organ of hearing distributed?
What two parts compose the external ear?
What membrane closes the auditory tube at its inner extremity?
What is the arrangement of the middle ear?
By what tube does it communicate with the back of the throat?
What is the labyrinth?
With what is it filled?
What three parts make up the labyrinth?
Of what parts does each consist?
What is the absolutely necessary portion of the organ of hearing?
Of which pair of cranial nerves is the auditory nerve a part?
What is the oval window?
What are the little bones of the ear called?
How many openings are there in the vestibule?
What earthy matter is contained in the vestibule?
In what bone of the head is the labyrinth placed?
What is the difference between the bony and the membranous labyrinth?
What is the perilymph? The endolymph?
With what canals does the vestibule communicate?
What are the hair-like projections in these canals?
What is the arrangement of the cochlea?
What are the two scalæ or staircases?
What is the auricle of the ear?
What is its duty?
Why do we apply the hand to the ear in order to hear more distinctly?
What effect have the long ears of animals?
How is the ear protected from injury?
What are the ceruminous glands?
What effect have the irregularities of surface of the external ear on the sound?
What is the chief feature of importance in the middle ear?
How do these little bones move?
What is the effect of the presence of air in the middle ear?
What is the chief muscle called connected with the membrane of the tympanum?
What is the action of this muscle and its effect on that membrane?
When the membrane vibrates, how is the sound transmitted to the internal ear?

How does sound reach the internal ear?

Do sounds always originate outside the body? What internal sounds may be heard?

Of what use is the fluid in the labyrinth?

What organ appreciates or judges of sound?

What parts must be perfect to ensure perfect hearing?

Are the vibrations of sound ever communicated through the bony walls of the head?

What conditions of sound itself are necessary to its appreciation?

What is the lowest limit at which sound can be heard?

What is the lowest and highest number of vibrations of sound that can be heard by the ear?

What does intensity of sound depend upon?

What does pitch depend upon?

What does quality of sound depend upon?

By which of these can we distinguish voices?

Does a sensation of sound end at once?

What facts apply to the eye and ear in common?

What peculiarities of the external ear exist in animals?

What changes in the semicircular canals, cochlea, etc. in animals?

What special bone have some animals in which the organ of hearing is lodged?

What peculiarities are found in birds? In reptiles? In fishes? In the mollusca? In insects?

VOICE.

The Organ of Voice.—The voice is produced by vibration of the vocal cords in the larynx while the air is passing through it. Speech differs from voice, being the use of sounds to express ideas. Speech is therefore an application of the voice. It is this power to convey thoughts by the voice that creates a wide distinction between man and other animals. The arrangement of organs for the production of the voice may be briefly said to consist of the larynx, the windpipe, the lungs, the mouth and nose, and the muscles concerned in breathing, for one and all are directly or indirectly interested in it.

The Larynx.—The larynx (Fig. 112) is situated in the front part of the neck, being more

·Fig. 112.—The Larynx (front view). 1, thyroid cartilage; 2, cricoid cartilage; 3, trachea or windpipe; 4, ligament.

prominent in the male, in whom it forms a very conspicuous object, commonly known as "Adam's apple." The mucous lining membrane of the mouth and throat

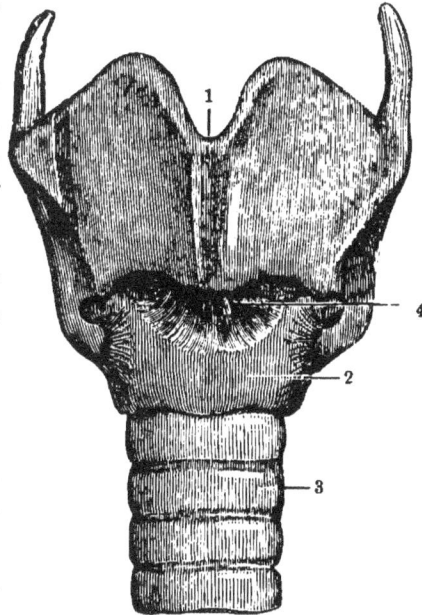

is continued downward into the larynx, and thence into the windpipe, which is a continuation of the larynx. The larynx is a tube made up of four cartilages, united together by ligaments, and having muscles attached to them. These bodies are movable upon one another, the object of which mobility will be presently explained.

The Vocal Cords.— The interior of the larynx is the part most directly interested in the production of voice. When we examine this portion of the organ we notice two clefts, or triangular spaces (Fig. 113). The upper one is bounded laterally by two folds of mucous membrane, called the upper, or *superior* or *false vocal cords;* the lower one by two folds, called the lower, or *inferior* or *true vocal cords.* Between these two spaces or cavities are the *ven'tricles* of

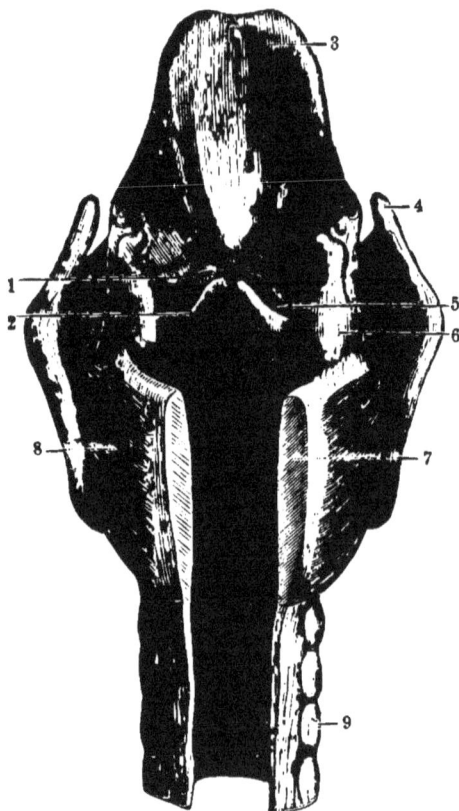

FIG. 113.—INTERIOR OF THE LARYNX.

1, upper or superior vocal cords; 2, lower or inferior vocal cords; 3, epiglottis; 4, thyroid cartilage; 5, vertical section of the glottis;.6, arytenoid cartilage; 7, section of cricoid cartilage; 8, muscle between cricoid and arytenoid cartilages; 9, trachea or windpipe.

the larynx, whose sides are mainly formed of muscles. The vocal cords are perfectly free to move, and are not interfered with in their action by any other structures. The vocal cords are only separated from one another about a third of an inch. The general plan of these spaces and cords will be readily understood by the accompanying diagram (Fig. 114).

Muscles of the Larynx.—The whole larynx may be moved at once, or only a part at a time. The muscles which are more directly concerned in the production of voice are those which are generally called the *intrin'sic muscles* of the larynx, which are so placed as to be capable of exerting some influence by their contraction or relaxation on this delicate vocal organ. The effect of their action is either to render the vocal cords or ligaments tense, or to separate the cartilages from one another,

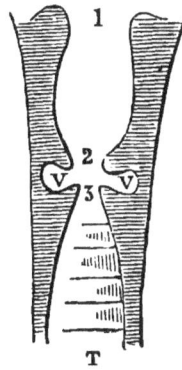

FIG. 114.—INTERIOR OF THE LARYNX (theoreti-' cal view).

1, folds between arytenoid and epiglottis; 2, superior or false vocal cords; 3, inferior or true vocal cords; V, V, ventricles of the larynx; T, trachea or windpipe.

so as to open the *glottis* (Fig. 113, 5), as the space of about a fifth of an inch in extent between the inferior vocal cords has been called, or to bring them together so as to close the glottis, or to relax the inferior or true vocal cords by drawing back the thyroid cartilage. Some of the muscles attached to the upper part of the larynx change the position of the epiglottis (Fig. 113, 3).

All these movements of parts are necessary to produce the changes of sound which are characteristic of the voice. The degree of tension of the vocal cords is also regulated by the action of muscles. As in all the special senses, the production of voice requires that the physical apparatus shall be perfect, and that the nervous supply to the organs shall be unimpaired. Nerves are distributed to the intrinsic muscles, and through them the muscular action in the production of voice is controlled and regulated. It can be readily seen that if these nerves should be injured or divided the action of the muscles would be checked, so that the conditions necessary to the production of voice could not exist.

Production of Voice.—In order to produce the voice it is necessary that the air should pass from the lungs through the windpipe into the larynx, cause a vibration of the vocal cords, and thence pass outward through the mouth and cavities of the nose. This is of course produced under the influence of the will, for otherwise voice would be produced at every breathing movement. The inferior vocal cords (Fig. 113, 2) are the parts of the larynx directly concerned in the production of the voice. It has been said that every other part of the larynx can be destroyed, and the voice may still be produced if these cords are left untouched. The appearance of the vocal cords at the moment of emission of the voice is represented in Fig. 116, A.

The passage of air through the glottis may be illustrated as follows: Take two pieces of bladder or India-rubber and stretch them over the open end of a tube, as represented in Fig. 115, so that each shall cover rather less than the opening, a space being left between them. If air be forced upward through such a tube by bellows a sound will result if the opening between the membranes be not too wide. The sound can be made to imitate the voice of an animal. If the membranes be made tighter or looser the sound will vary in character.

FIG. 115.—THEORY OF THE GLOTTIS.

Some writers on the larynx have described it as a wind instrument of the flute kind, or as being like a horn, the vibrations in those musical instruments being caused by a column of air passing through them; others have spoken of it as a stringed instrument, on account of the part taken by the vocal cords in the production of sound. Most writers, however, think it resembles more closely a reed instrument, like a clarionet. No matter what we may compare it to, the process of production of the voice is not at all simple. It includes a great variety of acts—the sending of air into the larynx, the contraction of the muscles of the larynx producing tension of the cords, their vibration, and the passage of air outward through the mouth.

The larynx is more developed in the male than in the female, and hence his voice is much stronger. Even when this organ is fully developed its action may be impaired by disease, although the disease may not be seated in the organ of voice. General debility may weaken the individual so much, as after severe sickness, that the muscles of the voice have not the power given to them to send the air through the larynx with such force as to cause the necessary amount of vibration of the cords.

Intensity and Pitch of Voice.—The *intensity* of the voice depends mainly on the force with which the air is driven through the larynx; in other words, on the pressure of the air. The number of vibrations of the cord in a given time produces the *pitch* of the voice. This is regulated by the degree of tension of the cords and the width of the space between them. When we wish to produce tones of high pitch we bring the edges of the cords more closely together, so that they become more tense (Fig. 116, c). When we wish to produce a low pitch the edges of the cords are separated (Fig. 116, b). By means of an instrument called a *laryn'goscope* (from two Greek words meaning "inspection of the larynx") which enables an observer to see, by means of a mirror, the interior of the larynx and the movements of its various parts during the production of sound, it has been noticed that before sound is produced the glottis is partly or completely closed by

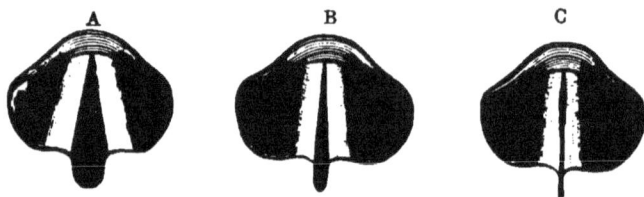

FIG. 116.—THE GLOTTIS AND VOCAL CORDS IN SINGING, AS SEEN WITH THE LARYNGOSCOPE.

A, opening of the glottis in the chest-voice and in ordinary inspiration; B, chest voice, for low notes especially; C, opening of the glottis in the falsetto voice or in high notes. The interval between the vibrating vocal cords is the glottis.

the bases of the arytenoid cartilages coming together, and the cords made tense, after which they are separated by the air passing between them and thrown into violent vibration. The full voice is not produced by the vocal cords alone, being intensified by the movement of the air in the cavity above it.

The number of vibrations of the cords on which the pitch depends is of course influenced by the length of the cords, their size, and the amount of tension. The longer the cords the lower the pitch; the tighter the cords the higher the pitch.

Quality of Voice.—It is difficult to explain the difference in quality of voice between individuals. Every person has his own quality, which belongs to him peculiarly and distinguishes him from every one else. The voice of the female is due to the fact that the larynx is more cartilaginous than in the male. The thickness of the walls, the size of the cavities, the form and length of the tube, probably affect the quality of voice. The presence or absence of the teeth and the shape of the tongue also have an influence. The change of voice in boys of about twelve to fourteen years of age is due to the fact that the ligaments and the opening of the larynx then become enlarged. In addition to the voice, as naturally effected, other sounds are capable of being produced from the vocal tube. Whispering is nothing more than articulation of the air that is breathed out of the vocal tube; whistling is caused by dividing the air as it passes through the lips.

The Singing and the Speaking Voice.—The human voice extends over a range of about three octaves in singing. The voice in singing differs from the ordinary voice in speaking by being made up of distinct musical tones, following in regular order in sequences, as they are called. The range of the singing voice is illustrated in Fig. 117, but this only shows the ordinary average. The quality and compass of voice vary greatly in different persons and in the two sexes. The deepest male voice is the bass, the highest the tenor,

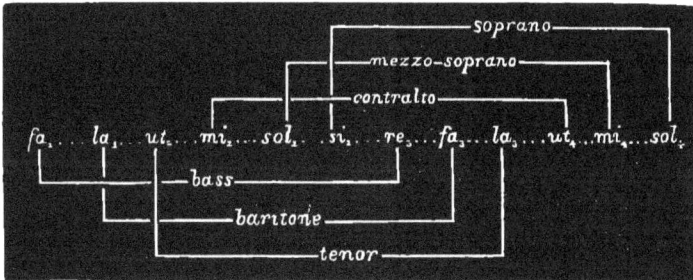

FIG. 117.—RANGE OF THE SINGING VOICE. (The figures denote the different octaves.)

and the baritone occupies a place between these two. The lowest female voice is the contralto or alto, the highest the soprano, and the intermediate voice the mezzo-soprano. When bass singers are capable of high notes the quality or tone is different from that of tenor or soprano singers. A

flute and a clarionet may both strike the same note, but the
tone or quality will greatly differ. The range of the different
forms of singing voice is well illustrated in the diagram
(Fig. 117). The singing voice is generally from the larynx
—chest-voice it is then called—but sometimes also from
the throat or pharynx. Cases are on record in which a
range of three and a half octaves was reached. The laryngeal voice is usually an octave higher in woman than in
man. The range of the human voice is generally considered to be three octaves, say from low fa or fa_1 (generally
known as F_1 on the scale of the piano), representing 174
vibrations a second, to high sol or sol_4, representing 1566
vibrations. The inferior or true vocal cords in men are
longer than those of women in the ratio of 3 to 2, the consequence of which is that the male voice is stronger and
lower in pitch. Exceptional voices have been known in
which there was wonderful range, as an Italian singer,
named Lucrezia Ajugari, to whom Mozart listened more
than a century ago, whose lowest note indicated 391 vibrations, the highest 4176. The notes produced in singing
are either chest notes, falsetto notes, or head notes. The
deepest notes are those from the chest, the highest from
the head.

Sounds are sometimes produced apparently from parts
deep down in the chest by the ventriloquist, as he has been
called. The word *ventril'oquism* is derived from two words
meaning "speaking from the stomach," and it would appear
possible that this might be true, judging from the depth
and distance of the sounds emitted. It is, however, nothing
more than a deception dependent upon the power of the individual to manage his voice as it proceeds from the larynx
and mouth, and to mislead his audience as to the direction
and quality of the sounds he utters. The different actions

and appearances of the larynx in the production of various voices is well shown in Fig. 116.

The strength of the voice depends upon the muscular movements regulating the action of the vocal cords, and not so much on the current of air passing through the larynx. These muscles are only about three quarters of an inch long, and all the different musical notes and tones are produced by the most delicate adjustment and variation of them. It is supposed that they can be accurately varied to a movement of 1-1200th to the 1-12000th of an inch.

The Formation of Language.—Speech, as already stated, is a series of articulate sounds to convey ideas, and is under the control of the will. Of course if a person has not the ideas to express, as is the case with the idiot, the presence of a perfect vocal apparatus will be of no use to him, as he is not capable of framing his ideas in words. Speech may be considered an evidence of the possession of intellect, and the quality of the intellect is often evidenced by the quality of the speech indulged in by its possessor. A parrot may speak distinctly, but its brain does not appreciate the meaning of what it says. Originally, before language was formed, words were probably derived from familiar sounds, such as the cries of wild beasts, the notes of birds, etc. The very names given to some of these sounds, as hissing, humming, snoring, grunting, whistling, wind, etc., are intended as nearly as possible to convey a notion of the sound itself. Words expressing such sounds in one language can be traced through a dozen other languages. The word "cuckoo," which is but the vocal expression of the cry of the bird, is said to be very much the same word in Greek, Latin, German, Arabic, Dutch, French, Persian, and a number of other languages. The clucking of the hen,

26

crowing, neighing, and bleating of various animals are reproduced in different languages. A "crash" is but an imitated sound, and so is a "thud," or dull, heavy sound or blow.

Vowels and Consonants.—The sounds that make up words are reduced to elementary ones, which we call *letters*, which placed together form *syllables*, and these again make up *words*. An *alphabet* is a collection of all the letters used in a language. The letters, or articulate sounds, are either vowels or consonants; the vowels being musical tones formed in the larynx, and passing out uninterrupted by the tongue or lips, although these occupy different positions in the production of the sound of each vowel. The strength of the sound is increased by the resonance of the air in the cavities of the mouth and pharynx. The name Aa (Danish for "river") has been given to several rivers, probably because as pronounced it represents the most easy and unimpeded flow of sound. Consonants are sounds formed in some part of the air-passages above the larynx, as the mouth, and increased in force by the action of the larynx. Some of them seem to come from the throat and base of the tongue, and are called *gut'tural* sounds, such as *ch* and *j*; those from the tongue and front part of the roof of the mouth are called *ling'uals*, as *s* and *sch*; those from the lips *la'bials*, as *b, p.* Some consonant sounds have a nasal quality imparted to them from a portion of the air passing into the passages of the nose—*m, n, ng,* for example. Some letters are called *explosives*, because they are uttered suddenly and cannot be sustained, as *b, d, g, p.* As these sounds do not often occur in Italian, that language is considered a much more agreeable one for the singer than either German or English.

The Voice and the Hearing.—We have already seen that

taste and smell depend upon each other for their perfection, and it is equally true that sight and touch are necessary to each other's full exercise. So, too, with the voice as connected with hearing. The vocal organs of a child may be absolutely perfect, but if hearing is wholly extinct from birth he cannot use articulate language, which includes the proper use of words to express ideas. He cannot, in other words, repeat sounds that he has never heard, and is therefore dumb as well as deaf. A person who has not a " musical ear "—that is, cannot appreciate harmony or melody in music—can never be an accurate singer, for he cannot reproduce sounds that he does not understand.

Vocal Apparatus of Animals.—The sounds emitted by animals are sometimes produced by their organs of respiration, but often through other means. Birds are endowed with very active respiratory organs, as we have already seen, and their continuous song is therefore much more easy than would be possible in almost all other animals. Some insects produce their familiar noises by the rapid motion or vibration of their wings—the mosquito, for instance. The friction against one another of the different hard parts of the cricket produces the peculiar shrill chirp which is so characteristic of that little animal. The grasshopper's shriek is produced by the animal rubbing its legs against its wings. It is said that in some parts of South America there exists a locust which has a kind of drum under its wings, the sound of which can be heard a mile or more : if a man of ordinary size had a voice in proportion it would be heard from one end of the world to the other. Those insects which fly most rapidly are the noisiest. The hiss of the tortoise and other reptiles and the croak of the frog are produced by a vibration where the windpipe opens into the pharynx.

In birds the important part of the vocal organs is at the lower part of the trachea or windpipe before it divides into the bronchial tubes. A sort of bony drum is present at this location in some birds, which is less simple in construction in birds that do not sing than in those which do. There is a small membrane of a crescentic or half-moon shape attached to a cross-bone, and when this membrane vibrates it produces the trill so familiar in the best singing-birds.

QUESTIONS.

How is the voice produced?

What is the difference between voice and speech?

What are the organs concerned in voice?

What is the prominence in the neck called in man?

What kind of a lining membrane has the larynx?

What is the larynx?

What is the appearance of the interior of the larynx?

What are the true vocal cords?

What are the ventricles of the larynx?

How far apart are the inferior vocal cords?

By what is the larynx moved?

What effect have the intrinsic muscles on the vocal cords?

What is the glottis?

What effect have the nerves in the larynx on the voice?

What action is necessary to the production of the voice?

Is voice voluntary?

What parts of the larynx are absolutely essential to the production of voice?

What kind of a musical instrument is the larynx?

What different actions does the production of voice include?

Is the larynx more developed in the male or the female?

What does intensity of the voice depend upon?

What does the pitch depend upon?

What different action takes place in the cords when we wish to produce high or low notes?

What is the laryngoscope?

How is the full voice produced?

On what does the number of vibrations of the cords depend?

What effect has a long cord on the pitch? A tight cord?
To what is the difference in voice of the sexes due?
What conditions affect the quality of the voice?
What change takes place in the larynx in boys?
What is whispering? What is whistling?
What is the deepest male voice called? The highest?
What intermediate voice is there in the male?
What is the highest female voice? The lowest?
What intermediate voice is there in the female?
What is the range of the singing voice?
How much higher is the laryngeal voice in woman than in man?
What is usually the lowest and the highest number of vibrations in the human voice?
What is the relative strength, etc. of the vocal cords in man and woman?
How are the notes produced in singing divided?
Which are the deepest notes?
What is ventriloquism?
What is speech?
What does strength of the voice depend upon?
What effect follows the minute vibrations of these muscles?
What defect of speech has the idiot?
How was language first formed?
On what principle are words like "hissing," "snoring," "grunting," "cuckoo," etc. derived?
What are elementary sounds called that make up words?
What is an alphabet?
What two divisions are there of letters?
What are vowels? How are they formed? How are they increased in strength?
What are consonants? How are they formed?
Into what divisions are consonants divided?
How are nasal sounds produced?
What are the explosive consonants?
How are the voice and hearing related to one another?
Why is a deaf person usually dumb also?
How are the sounds or voices of animals produced?
How do insects produce their peculiar sounds?
How is the noise of the cricket produced?
How is the sound of the grasshopper caused?
How is the croaking of the frog produced?
What is the anatomical arrangement in the windpipe of birds for singing?

Hy'giene may be defined as the art of preserving health, whether of the individual or of communities. It is properly studied in connection with physiology, for it would be impossible for one not having a correct knowledge of the form and structure of the various organs of the body to understand the laws and precautions necessary for their preservation in a state of full health. Human life would be greatly prolonged if all the world were educated to avoid the causes which so often lead to sickness and death ; and life, under such a knowledge of preventible disease, would be more enjoyable and more vigorous, for good health is the chief basis of such enjoyment and vigor. *Disease* is a derangement of healthy physiological action, and, being a result of the violation of the laws of health, it becomes necessary to put an end to such violations of law, if it be possible to reach them, before a perfect cure of the disease can be effected.

In studying the physiology of the various organs and functions (pp. 1 to 304) a large amount of information usually taught under the subject of hygiene was imparted, for the two branches are practically inseparable. The study of hygiene may, however, be extended to include a vast number of other subjects not usually or necessarily taught in connection with physiology proper. Prominent among these is a study of avoidable causes of disease, espe-

306

cially as regards the habitual use of intoxicating liquors
and of substances injurious to health.

Hygiene of the Bones.—The healthy condition of the
bones is maintained by some of the same influences that
preserve the general health of the individual. Thus, if
children are restricted to a diet which is deficient in the
materials that go to the formation of bones—such as phosphate of lime—the firmness and solidity characteristic of
these structures will be wanting. A disease called *rickets*
is sometimes caused in this way, the bones being weak and
imperfectly formed, and in bad cases almost as yielding as
if made of wax. The bones of the leg, for example, give
way and become bowed, being unable to support the
weight of the rest of the body. Such diseases are sometimes avoidable by attention to the laws of health. Of
course an excess of phosphate of lime would have an opposite effect, and would render the bones brittle and liable to
be broken from slight causes.

Exercise and fresh air and the avoidance of over-pres-
sure and of faulty positions of the limbs will aid in preserving the health of the bones even when there is a
tendency to disease in them. The sitting and standing
attitudes should be easy and graceful, so as to avoid habits
that may injuriously affect the shape and position of the
various bones. Tight lacing should be especially avoided
as a cause of unduly compressing the bones of the chest,
thus altering the shape so that the lungs do not have full
room for expansion. Broken bones should be carefully
placed in as natural and easy a position as possible, and so
maintained until surgical assistance can be called upon.

Hygiene of the Teeth.—Attention to the preservation of
the teeth by the individual himself will in many cases not
only tend to maintain them in healthful vigor, but save

him from much pain and distress in after-years. The teeth
are injured by want of cleanliness, by changes in temperature, by the action of acids, by mechanical violence—such as
cracking hard substances between the teeth—by extremely hot or extremely cold articles of food, etc. A hard
deposit of "tartar" may occur upon the teeth from the
saliva, sometimes forming a crust requiring the dentist's
aid for its removal; but its formation may be prevented
by the daily use of a tooth-brush and water. The enamel
of the teeth, as the outer portion is called, is sometimes
broken or injured, and the dentine soon becomes decayed,
leading to caries of the teeth, as it is called, and the formation of a cavity, which, soon involving the nerves supplied to the teeth, causes the sensation of pain. The
dentist's art relieves this by the application of gold or
other material after the cavity has been thoroughly cleansed
of all decaying materials.

The **Hygiene of the Digestive Organs** has already been
fully discussed in connection with the Physiology of Digestion. (See pages 49–108.) The effect of alcohol and narcotics upon the digestive organs will be referred to in a
subsequent section. (See pages 317 and 319.)

Hygiene of the Respiration.—As healthy respiration is
dependent upon an atmosphere of healthful quality and
abundance, the first element is to preserve the air we
breathe in its purity. A sufficiency of oxygen in the composition of the air is the chief desideratum, and it is the insufficiency of this element, in addition to the presence of
impurities, that constitutes the main contamination of the
breathing-air in the interiors of houses, in halls, school-
rooms, etc. This condition is caused in various ways, but
mainly because a free ingress of the purer air outside is
checked in some manner, as by tightly-closed windows or

doors, which thus also prevent the escape of the impure air from within, and where several persons are in the same room by the accumulation of gaseous matter thrown off from their lungs or bodies. Where many persons are assembled together, and no adequate outlet is offered through windows or doors for the escape of such gaseous matter from the room, the place is said to be badly *ventilated*, and the foul air continues to be breathed and rebreathed until the sensation of drowsiness, headache, and fatigue becomes almost overpowering.

If such be the result of long continuance in an atmosphere breathed by persons generally inferred to be healthy, how much more serious it becomes if we are continually exposed to the injurious effects of breathing an atmosphere loaded with the exhalations from the lungs or bodies of those affected with disease! The impurities thus emanating from diseased persons are not only unfit to be rebreathed, but they may give rise to serious diseases. For purposes of ventilation and the healthy exercise of the organs of respiration, the rooms occupied by both sick and well should be properly ventilated by windows that will admit light and air during the day and keep the atmosphere pure at night. If fires be employed in sleeping-rooms, there must be sufficient outlets provided for the escape of the overheated air. Openings at the top of the room will greatly add to the preservation of the purity of the air by allowing hurtful gases to escape and by admitting the entrance of fresh air, from without. The impure air at the bottom of the room should be carried off, by shafts into the flues or chimneys, into the current passing upward and outward to the external air.

Allusion is made elsewhere to the effect of tight lacing upon the organs of the chest and abdomen, by which the cavity of the former is deformed and diminished in size and the free movement of the lungs interfered with. The stomach and liver are also pressed downward and in their unnatural position interrupted in the proper performance of their duties. The healthy lungs cannot be thus impeded in their movements of expansion, and the air prevented from entering them in sufficient quantity, without serious consequences ensuing. The lower portion of the chest is naturally the largest (as seen in Figs. 1 and 49, pages 22, 131) and well adapted for the enlargement of the capacity of the lungs in breathing, but tight lacing reverses the order of things and makes this the narrowest part of the chest.

In breathing, the air should pass in and out through the nose, not through the mouth, and at night the mouth should be kept closed. Attention should always be paid to the attitude, a standing or sitting position in which the chest is cramped being especially to be avoided. If the shoulders be maintained erect and a habit of stooping studiously shunned, and full inspirations and expirations be taken, the lungs will act to their full capacity and respiration will be perfectly accomplished.

The process of healthy respiration is too often interfered with by the action of poisonous substances taken into the stomach, as alcohol, etc., or inhaled by smoking, as tobacco; but these will be referred to more fully in another section. (See pages 318 and 325.)

Hygiene of the Circulation.—The even course of the circulation is greatly assisted by exercise, by proper attention to the general health, and by other causes. Rapid exercise, moderately employed, contracts the muscles, increases the flow of blood in the vessels, and quickens the action of the heart and of the circulation generally, while it

sharpens the appetite and promotes the process of digestion. Any articles of wearing-apparel that press too
tightly upon the bloodvessels on the surface of the body
impede or arrest the circulation, and thus injuriously affect
also the temperature of the parts beyond the point affected.
Changes of temperature (as elsewhere shown), wet feet, and
similar conditions resulting from exposure, affect the local
circulation, and secondarily the general circulation in other
and more distant parts of the body.

The action on the organs of the circulation of various
substances taken as a habit, such as alcohol, tobacco, etc.,
will be considered in a subsequent section. (Pp. 316-17.)

Hygiene of the Skin.—The importance of the skin as a
medium for preserving the general health or of producing
disease cannot be too forcibly dwelt upon. Fortunately,
the skin is not very impressible to outside agents unless
applied directly to it, as by force, by rubbing in, or by
immediate contact. When the surface of the body becomes affected, however, by disease, the general system
often sympathizes with it. The health of the whole body
is improved and maintained by attention to two very
important but essential external agencies—clothing and
bathing.

The *clothing* must be clean, and must not be worn so
long as to be soiled or contaminated with the waste matter
given off from the surface of the body. Under such circumstances the pores become closed, and the matters usually
given off from the blood through the skin do not have a
channel through which they can pass off from the body,
but remain in the system to exert an injurious and, to some
extent, poisonous effect. Clean and frequently changed
underclothing, as it comes directly in contact with the
skin, is especially desirable and important, particularly as

it readily absorbs the moisture given off from the surface. It will be seen how important it is that clothes worn during the day and the bedclothes used during the night should be well aired, and thus purified.

We have already seen, in studying the subject of Animal Heat (p. 179), that the temperature of the body remains about the same under all favorable circumstances. Clothing is necessary to prevent too rapid loss of heat, either from contact of the surface with cold air or from the evaporation of the perspiration ; and the materials employed depend, not on the warmth they themselves supply, but on the fact of their being non-conductors of heat. Porous materials are usually employed for this purpose, as the pores or openings are filled with air and their conducting power is low. Wool, silk, cotton, and linen are the articles generally used.

The material of which the clothing is made is important as an element in preserving health. Woollen clothing is employed because, being a poor conductor of heat, it does not allow the natural warmth of the body to escape through it, and it absorbs the perspiration without becoming moistened by it. For these reasons flannel is worn, or should be worn, next to the skin at all periods of the year, even, to a slight degree, during the season of extreme heat. In milder climates cotton may be worn ; it readily absorbs moisture. Linen is a cool material for wear in summer, as it is a good conductor of heat and rapidly carries off the warmth of the surface. Silk is a good non-conductor of heat and absorbs very little moisture.

When the skin is rapidly chilled by exposure of any kind, what is popularly known as " taking cold " results— a condition dependent upon the blood at the surface tempo-

rarily receiving an atmospheric shock, by which the blood
is sent in unusual quantity to internal parts, which, being
susceptible, become congested and inflamed, producing
sore throat, inflammation of the lungs, or some other serious condition of important organs. Warm applications
and the use of such internal medicines, under proper
advice, as will stimulate the skin to increased secretion
and the blood to more active circulation in the direction
of the surface of the body, will frequently relieve the condition of " cold " thus induced.

Bathing.—Bathing the surface of the body, accompa-
nied with the proper amount of friction from the towel, is
necessary and beneficial by removing any dirt or dust that
may have adhered to the skin, and by clearing it of an
excess of matter that may have been left as a deposit after
the evaporation of the watery particles poured out by the
sweat-glands and the secretions from the sebaceous glands,
which in our study of the Physiology of the Skin we
found to be so liberally distributed to it. Bathing has its
rules, however, which cannot be gainsayed without injury
to the general health. A cool bath taken daily will, as a
rule, agree with those of strong, healthy constitutions,
but it should be immediately followed with an active
and decided friction of the skin with a coarse towel
until a glow is produced, and a certain amount of after-
exercise must be practised. A chill after a cold bath
is evidence that it does not agree with the bather. Those
of more feeble health should not bathe so frequently, but
should rather use tepid or moderately warm water say
once or twice a week, and for a short time only ; but the
skin should be wiped thoroughly dry and then rubbed
until the surface becomes warm, Sea-bathing has a still
more invigorating effect,

27

Too often perfumery is employed to disguise the odors that accompany an unclean condition of the body, and cosmetics are used to give an artificial tone and color to the face; but many of the latter are made up of poisonous ingredients, such as lead, and few of them are perfectly harmless. Soap would be decidedly beneficial for the removal of dirt from the surface, and perfumes, if employed at all, should follow instead of preceding its use.

The simplest forms of baths, classified according to temperature, are the following:

Bath.	Water.	Vapor.	Air.
Cold	33° to 65°		
Cool	65° to 75°		
Temperate . . .	75° to 85°		
Tepid	85° to 92°	90° to 100°	96° to 106°
Warm	92° to 98°	100° to 115°	106° to 120°
Hot	98° to 112°	115° to 140°	120° to 180°

Bathing should not be indulged in after a full meal, about three hours being allowed to elapse, so that sufficient time may be given for digestion to be accomplished. Early in the morning is regarded as the best time for taking a cold bath, but any one unaccustomed to such a bath should begin at a higher temperature—say 65° or 70°— and gradually reduce it to a much lower point. A stay in the bath of ten to twenty minutes is sufficient for all healthful purposes, and even at the seashore it should not be protracted to a much longer period. The body should not be chilly when entering into the water. History tells us that the ancient Romans indulged in numerous daily ablutions, especially in the form of the vapor-bath, afterward anointing themselves with oil, while some of them exposed their bodies to the bright rays of the sun, in the belief that under the influence of this *sun-bath* they would be physically developed and strengthened, as plants are when similarly exposed.

The Action of Alcohol and Narcotics.—The earliest employment of alcohol as a beverage is said to have been in the days of Noah, who planted a vineyard and became intoxicated on the wine made from the fruit thereof. The common source of such drinks in all times has been the grape, but the palm tree, the pomegranate, melon, and other fruits, and rice, have been employed for this purpose, the article being subjected to a process called fermentation; for alcohol itself does not exist as such until the substance containing sugar is decomposed in the presence of water. The elements of which the sugar is composed —hydrogen, carbon, and oxygen—become rearranged to form carbonic acid, alcohol, volatile oils, and ethers.

It is said that alcohol was called *spirit of wine* because, in olden times, it was found that if wine were kept at a boiling-point for a few minutes the stimulating principle departed from it like an invisible spirit, and the wine was no longer intoxicating. Wine is usually a product of the fermentation of grapes; cider, of the fermentation of apple-juice. Alcohol is a clear, colorless, inflammable liquid lighter than water; it freezes at very low temperatures, and is therefore used in thermometers when such temperatures are to be taken.

Is Alcohol a Food?—Alcohol has been considered by some writers as a food, but, on whatever theory such a view is based, the results of its use are very different from those observed after taking nourishment. Alcohol creates thirst and by its affinity for water abstracts fluid from the tissues, is not changed when it enters the circulating blood, lowers the animal temperature, and does not relieve hunger, and differs in all these respects from food taken for the nutrition of the body. It was formerly the custom to give grog to soldiers and sailors as part of their daily supply of rations, but all experiments have shown that the army and navy, in every part of the world, were more effective when this ration was entirely withdrawn. The alcohol does not itself supply nourishment, and does not in any way aid in the digestion of other articles of food. In countries characterized by extreme cold, alcohol does not maintain the heat of the body, which requires the taking of oily animal substances to produce the effect. Death, indeed, frequently occurs in very cold climates from even moderate use of such stimulants.

Does Alcohol Produce Warmth ?—It is even a delusion that alcohol produces a general sensation of warmth through the system. The flushed face of the habitual drinker does not indicate any such elevation of temperature, but rather a deadened sensibility of the small nerves of the face which control the contraction and dilatation of the minute bloodvessels; so that the blood fills up their interior and gives rise to the appearance of fulness so often seen in such individuals. Intemperate habits sometimes keep these vessels permanently distended, producing an eruption not easily gotten rid of.

Adulteration of Alcoholic Drinks.—If all these statements be true of alcohol in its purity, how much greater the evil attendant upon its use in the impure forms in which it is taken daily by thousands upon thousands of thoughtless or intemperate people, adulterated with ingredients that are positively injurious, perhaps absolutely poisonous! Wherever government inspections have been made of importations of alcoholic liquors, the amount of adulteration detected has been found to be perfectly astounding. Experiments with alcohol in its pure state have been made at various times on the lower forms of animal and vegetable life, and have conclusively proven that it is absolutely destructive to vitality.

Action of Alcohol on the Digestive Organs.—Alcohol acts as an irritant to the stomach and digestive organs, producing at first loss of appetite and dyspepsia; and if a habit of drinking it be persevered in, the structure of the stomach may undergo alteration, becoming thickened, and perhaps inflamed. (See page 319 for its action on the liver.)

Action of Alcohol on the Heart.—Upon the heart alcohol at first acts as a stimulant to its action, producing an increase in the number and force of the pulsations; so that the period of rest to which we referred in studying the Physiology of the Circulation (page 158) is diminished, and consequently a greater amount of strain and labor is placed upon an organ which needs regular and constant rest in its frequently-recurring periods of active duty. If the work performed by the heart every day be estimated— as it has been by some writers—as equivalent to the lifting of so many tons one foot, the additional work imposed upon it by the internal use of alcohol would represent an excess of from twenty to twenty-five tons in the course of forty-eight hours. If the number of beats be increased only five every minute, and this continued stimulation should become a habit, the heart would beat over seven thousand times a day more than its accustomed number, entailing upon it so much additional labor.

Although the action of the heart is thus quickened, it soon becomes enfeebled, and requires additional quantities of the same stimulant to revive it. The circulation, therefore, becomes irregular, being sometimes excited and at other times fatigued. The heart may soon beat intermittently—that is, with irregular intervals between the

27 *

pulsations—or it may become dilated from over-action and the valves lose their rigidity, with thickening of the membranes lining the cavities and degeneration of the large and small bloodvessels connected with the heart. We can readily see how easily the heart may suddenly fail to do its work or the vessels become ruptured, and the blood that should flow continuously and uninterruptedly through the heart be thrown back upon other important organs, as the brain or lungs, producing apoplexy of the former or serious obstruction in the latter.

Action of Alcohol upon the Respiration.—The influence of this poison upon the lungs is noticeable in its interference with the proper oxygenation of the venous blood. We have seen, in studying the Physiology of Respiration, that this great function includes the conversion of venous into arterial blood in the lungs through the influence of the oxygen breathed in at every inspiration. If the carbonic acid given off by this process be prevented from having a ready outlet, its poisonous effects upon the system will soon be apparent. As alcohol has been conclusively shown to be a local irritant, its passage through the walls of the air-cells must be very injurious to such delicate structures. Alcohol changes the shape and appearance of the red corpuscles, causing them to be irregular and shrunken, with notching of their edges. Habitual drinking of weak alcoholic liquors has the effect of thinning the blood, so that it escapes too easily from the vessels and makes its appearance near the surface of the body, giving the skin a swollen or bluish hue; or if strong drink be habitually indulged in, coagulation of the blood may occur and clog up the smaller vessels or the great organ of the circulation, the heart itself, and cause instant death.

The effect of alcohol on the lungs themselves is soon visible. The relaxed bloodvessels found in these delicate structures become easily congested when paralyzed by this poison, and death often results from such a condition. Even those who seem capable for a while of taking alcoholic liquors in moderation without apparent danger may suddenly be seized with disease of the lungs attended with cough and pain, and rapid consumption may follow.

Action of Alcohol upon the Liver.—The evil effects of alcoholic drinks are exhibited in every important organ of the body, but especially in the liver, brain, and heart. In our study of the Physiology of Digestion (pages 65–69), we showed how alcohol and other thin liquids passed at once into the veins of the stomach, and thence were carried directly into the portal vein to the liver before being distributed to the other portions of the body. It is through this *portal circulation,* as it is called, that the great injury inflicted upon the liver is effected by alcoholic liquors taken into the stomach. They are not digested, and do not wait, as thicker fluids do—such as soups, which are really nutritious—to be acted upon by the stomach at its leisure, but are rapidly absorbed, as we have just stated, and spend their hurtful force at once upon the liver. There is a remarkable change in the appearance of the liver in a person who has been drinking liquors to excess. This condition is called *cirrho'sis,* from a Greek word meaning "yellowness," but it is also known as "gin-drinker's liver" and "hob-nail liver." If habitual drinkers could be made familiar with the contrast between the appearance of the healthy liver and of the same organ when so seriously affected by the influence of alcohol, the effect might be that of a useful temperance lesson.

This is what is known as an *organic disease*—that is,

the structure of the organ is entirely changed; while a *functional disease* is one which affects the organ in the performance of its duties without necessarily altering its structure. Another instance of such an organic change is seen in the fatty degeneration in certain vessels of the body consequent upon the use of alcohol to excess, and in its worst form in fatty degeneration of the heart itself. An organic disease of the liver, such as cirrhosis, just referred to, would clog up all the important vessels of that organ, interfering with the proper formation of the bile and with its passage outward through the bile-ducts, and destroying the substance of the liver, so that it could no longer perform its proper duties, and finally producing death.

Action of Alcohol upon the Brain.—The effect upon the brain of small quantities of alcohol taken occasionally is to increase the amount of blood going to that organ. This is still more marked with the increase of quantity and the taking of alcohol as a habit. It has been observed in the brains of those who have died from *alcoholism*—as the condition resulting from the abuse of intoxicating liquors is called—that the coats of the bloodvessels have become weakened, so that they sometimes give way and produce what is known as apoplexy or effusion of blood into or upon the brain-substance. Independent of this, however, the brain-substance is found also to be hardened and wasted away and shrunken, with effusions of water into the various cavities. The effect upon the mind of the individual is, of course, disastrous: it does not receive correct impressions; it becomes clouded and confused, with delusions of various kinds; it does not have rest; there is loss of sleep, with failure of the reasoning faculties and general unsteadiness of the muscular system; to which is added intense desire for drink, called *dipsoma'nia,* or " mania for drink."

Under such a perverted condition of the mental faculties, the moral powers become blunted and impaired, and
crime and poverty soon follow in the train. When once
formed, the habit controls the man, not the man the habit;
the individual becomes in every sense a moral, mental, and
physical wreck, from which death is often a relief to himself and his friends. Statistics carefully made in various
portions of the world have conclusively proven that crime
keeps pace with drunkenness, and that lawless deeds of
every description occur much more frequently in communities in which alcoholic liquors are indulged in to excess.
The experience of the court-room in all parts of the
world has abundantly shown that drink is the exciting
cause of crime in a vast number of cases.

In its effects upon the brain and nervous system generally alcohol belongs to the class of *narcot'ics* (from a Greek
word meaning "stupor"). Its first effects are stimulant,
and seem to augment the activity of the muscular system
and to excite increased warmth and mental activity, but
such effects are transient and deceptive; and if the habit
be continued, self-control is lost and the poisonous qualities of the liquor are allowed full play, the individual
being conscious of the injury to his system thus self-in-
flicted, but not having the power to check himself in the
unlimited indulgence. In due time the brain becomes the
seat of all kinds of delusions; sounds are heard that have
no reality, objects are seen which have no existence, and
a condition known as *delirium tre'mens* or *mania-a-po'tu*
("mania from drink") results. Sometimes paralysis or loss
of power of both mind and body occurs, from the action
of the poison upon the nervous matter and upon the membranes covering the brain and spinal cord.

Action of Alcohol on the Temperature.—If a delicate
v

thermometer be placed under the tongue of a person who has taken a quantity of alcoholic drink, it will be found that, though there may at first be a slight increase in the temperature, there will soon be a reduction in the natural heat of two or three degrees, lasting for several hours.

Inheritance of Alcoholism.—One of the surgeons of the Salpêtrière Hospital, Paris, has for a number of years studied the nervous affections of children, especially among the children of alcohol drinkers. He found in 83 families in which there was nervous excitability traceable to alcoholic origin 410 children. Of these, 108 had convulsions; and in the year 1874, 169 were dead and 84 of the 241 still alive were subject to epilepsy.

Mortality from Alcohol.—A recent writer states that there are no less than twenty diseases acquired by mankind from the use of alcoholic liquors, and in the list he does not include such diseases as cataract and amauro'sis, or loss of power of the optic nerve, which are sometimes traceable to the same cause. Premature decay and old age might be added to the list. The mortality from alcohol is very great, as the diseases caused by its use attack those whose constitutions are already enfeebled by drink and are often beyond the reach of medical aid. It has been calculated by a distinguished writer that in the United Kingdom of Great Britain and Ireland at least a hundred deaths occur every week from alcoholic excitement, and a thousand deaths per week from the diseases which, directly or indirectly, follow from indulgence in the use of alcohol.

Action of Opium upon the System generally.—Opium is a poison of the narcotic class which when properly given, under medical advice, is capable of acting as a valuable remedy in relieving pain, producing sleep, etc.; but when it is used unrestrictedly and for self-indulgence, a habit is established which will eventually wreck the unhappy individual who indulges in it. Opium is the thickened juice of the poppy. Morphia is its active principle, and laudanum, paregoric and Dover's powder, and other similar preparations, owe their virtue for relieving pain and causing sleep to the opium which is the active ingredient in each. Although given in some of these forms to very young children to produce sleep, opium is most injurious to the earliest periods of life, dangerous symptoms sometimes occurring from the giving of even the smallest doses—a drop of laudanum, for instance. "Drops," "soothing syrups," etc., are given and taken recklessly in this way, often with fatal effects.

We frequently hear of opium-eating, the "opium-habit," the "morphia-habit," etc.—all synonymous terms for the uncontrolled, and ofttimes uncontrollable, indulgence in the use of the same narcotic. The effect of opium is not restricted to the nervous system : the digestive organs become impaired in their activity, the countenance is shrunken and sallow, and the powers of the intellect are dulled and weakened. The delusion that healthy natural sleep can be produced by opium or other narcotics should be at once dispelled : the individual does not awaken to a day of rest, as after a night's refreshing sleep, but to a disturbed and unsettled feeling of unrest of a most distressing kind. The same results follow the use of opium whether it be smoked or eaten, except that a larger quantity will be taken in the latter way. Paralysis frequently occurs as a result of its employment.

Action of Chloral Hydrate, Chloroform, etc.—Chloral is one of the new remedies of the past quarter of a century. It is made from alcohol by a peculiar chemical action, is a valuable sedative remedy when properly given under good medical advice, but, like opium, may be abused by self-indulgence, until the chloral-habit—*chlor'alism*, it has been called—becomes as firmly engrafted upon the system of the individual as the opium-habit. It soon impairs the digestive process, deranges the circulation, disturbs the action of the heart and nervous system, and, while it may have originally promoted sleep when taken for that purpose, experience will soon show, in those addicted to the habit, that sleep cannot be procured without it. By chemical action it also makes the blood much more fluid.

Chlor'oform, as is well known, has been largely employed by inhalation for the relief of pain in surgical operations, but even in the hands of skilful surgeons has not been considered free of danger to life. The habit of resorting to it by others for the relief of pain, without taking advice as to its use, is very much to be condemned; for no rules can be laid down for its employment under such circumstances that will prevent the possibility of the individual, at some time or other, taking the few additional inhalations that will be alone necessary to terminate his pain and his life at the same moment. Even if habitually inhaled with safety to life, the digestive organs and nervous system soon become disordered, and the action of the heart is weakened.

Indian hemp is sometimes taken as an intoxicating substance, on account of the drowsiness and visionary dreams associated with its use; but the effect of all such articles, taken as a habit, is to excite and cloud the intellect, and their employment often leads to insanity.

Action of Tobacco.—Tobacco also belongs to the class of poisons, and chiefly owes its injurious effects to the presence of an active principle in it called *nicot'ine*. Young persons who use it for the first time are often attacked with sickness of the stomach and vomiting, and are apt to suppose that if they can master this obstacle to its further employment all danger is past, and that thenceforth there will be only pleasant indulgence; but such an idea should be dispelled. When the habit is thoroughly formed, it becomes, like that of alcohol-drinking, uncontrollable, and the poisonous effects of the tobacco are evidenced by nervousness, debility, disturbance of the stomach, vomiting, giddiness, and other similar symptoms. Convulsions and an irritable, weakened state of the heart have also been frequently noticed as accompanying the excessive smoking or chewing of this narcotic.

Experience in the United States Naval Academy, at Annapolis, of the ill-effects of tobacco, as evinced in the general unsteadiness of those indulging in it, led to a law forbidding its use in that institution, and similar action was taken at the Military Academy, at West Point. The results have been such as to satisfy all parties that the health of the cadets has been sensibly improved by such restriction.

As was remarked of alcoholic liquor, the article itself is poisonous when perfectly pure; and if adulterated—as tobacco undoubtedly is in the form of cigarettes, snuff, cigars or chewing tobacco, sometimes with other narcotics, as opium —the injurious effects on the nervous system must be greatly increased. In addition to the nicotine in the tobacco, smoking also develops carbonic oxide, which is an active poison, itself producing, when inhaled with the fumes of the tobacco, drowsiness, irregularity of the heart's action and sickness of the stomach.

Tobacco-smoking injures the red corpuscles of the blood and greatly disturbs the action of the heart and bloodves-

28

sels. When indulged in to any great extent, it produces confusion of sounds in the ear, with ringing in that organ and irritability and imperfection of the vision, which sometimes amounts to total blindness, and then receives the name of *tobacco amauro'sis.*

In some observations recently made in France it was found that the mean temperature of the body for the twenty-four hours in non-smokers of average constitutions was about 98° Fahr., while in those addicted to the use of tobacco the mean temperature was 98.6° Fahr. In those of weak constitutions the temperature rose to a much higher degree. Tobacco, therefore, may be said to raise the temperature of the body one degree, on the average.

Observations similarly made show that the mean pulse-rate is 72.9 among non-smokers of average constitutions, and in those addicted to the use of tobacco 89.9—an increase of about seventeen pulsations of the heart every minute. We may illustrate this more clearly by stating that to every thousand pulsations in one who does not smoke there would be one thousand two hundred and thirty-three pulsations in him who does smoke. The effect of such increased action of the heart must be very injurious, giving it increased labor, and augmenting the number of beats of the heart more than twenty thousand every day.

QUESTIONS.

What is hygiene?   What relation has it to physiology?
What is disease?
How is the healthy condition of the bones maintained?
What effect has diet upon the bones?
What is the effect of tight lacing on the bones?
What influences affect the health of the teeth?
What is caries of the teeth?
What is healthy respiration dependent upon?
What are the causes of impurity in the breathing air?
What is bad ventilation dependent upon?
How is it to be remedied?
How does tight lacing affect the respiration?
How does position affect the respiration?
What effect has exercise on the circulation?   What effect have clothing and temperature?
What are the two important agencies for preserving the health of the skin?
What effect has inattention to cleanliness upon the skin?
What is the object of clothing?
What materials are employed for this purpose?   And why?
What is meant by "taking cold"?
What is the effect of bathing?
When should a cool or a tepid bath be employed?   ⸰
What is the objection to using cosmetics?
What is the usual temperature of a cold bath?   Of a hot bath?   Of a hot vapor-bath?
What are the best times for taking a bath?
What is said to be the earliest employment of alcohol?
What are the common sources of alcohol?
How is alcohol formed, chemically?
What are the physical properties of alcohol?
What is wine?   Cider?
Why cannot alcohol be considered as a food?
What effect has alcohol in cold climates?
Why does alcohol seem to create warmth?
How does alcohol affect the stomach?   The heart?
Does it increase or diminish the work of the heart?
How many more beats a day does alcohol produce?

What effect has alcohol upon the respiration? Upon the red corpuscles of the blood? Upon the blood itself? Upon the structure of the lungs?

What effect has alcohol upon the liver? Why does it act at once upon the liver?

What is the condition of the liver called which results from drinking alcohol to excess?

What is the difference between an organic disease and a functional disease?

What effect has alcohol upon the brain? Upon the mind?

What is alcoholism? Dipsomania? Delirium tremens?

What is a narcotic?

What effect has alcohol on the temperature of the body?

What is opium? Its active principle?

What is the opium habit?

What is the effect of opium when taken as a habit? Of chloral? Of chloroform?

What is chloralism?

What is the active principle of tobacco?

What is the action of tobacco, taken as a habit?

What other poison is developed in tobacco-smoking?

What effect has tobacco on the vision? On the temperature? On the pulse?

INDEX.

329

www.ingramcontent.com/pod-product-compliance
Lightning Source LLC
Chambersburg PA
CBHW021457210326

41599CB00012B/1041